Physiology of Membrane Fluidity

Volume I

Editor

Meir Shinitzky, Ph.D.
Professor
Department of Membrane Research
The Weizmann Institute of Science
Rehovot, Israel

CRC Press, Inc.
Boca Raton, Florida

Library of Congress Cataloging in Publication Data
Main entry under title:

Physiology of membrane fluidity.

   Bibliography: p.
   Includes indexes.
   1. Membranes (Biology)--Mechanical properties.
2. Membranes (Biology)   I. Shinitzky, Meir, 1939-
QH601.P49   1984   574.87′5   83-24077
ISBN 0-8493-6141-9 (v. 1)
ISBN 0-8493-6142-7 (v. 2)

   Direct all inquiries to CRC Press, Inc., 2000 Corporate Blvd., N.W., Boca Raton, Florida, 33431.

© 1984 by CRC Press, Inc.

International Standard Book Number 0-8493-6141-9 (v. 1)
International Standard Book Number 0-8493-6142-7 (v. 2)

Library of Congress Card Number 83-24077
Printed in the United States

FOREWORD

Early in evolution, life processes were confined by membranes to separate compartments in order to minimize random dissipation of energy and information. In membrane-sealed organelles, reactions are conducted by speciallized macromolecules (e.g., enzymes) at a very high local concentration, while small substrates and products diffuse selectively through the membrane boundary. Biochemical reactions and signal reception on the membrane surface itself take place in a two-dimensional array where diffusion is nonrandom and local concentration can be extremely high. In this type of organization, the membrane lipid fluidity plays a key regulatory role. This complex parameter, which combines structural and diffusion aspects of the membrane lipid domain, is the principal issue of this volume. The main intention in publishing it is to provide, for the first time, a comprehensive account on the role of lipid fluidity in physiological processes and on the mechanisms which are involved in natural and pathological alterations of membrane fluidity.

Upon studying it, the reader may realize that at the current state-of-the-art membrane fluidity can be modulated both in vivo and in vitro through metabolic pathways or passive exchange. It may be expected that manipulation of membrane activities through changes in lipid fluidity will soon become a major tool for amplification of specific functions, as well as for rectification of abnormal activities. My sincere hope is that this volume may provide a solid base for the promotion of such approaches.

Meir Shinitzky

THE EDITOR

Meir Shinitzky, Ph.D., is a Professor of Biophysics in the Department of Membrane Research, The Weizmann Institute, Rehovot, Israel.

Dr. Shinitzky received his Ph.D. degree from this institute in 1968, and thereafter, spent two years at the University of Illinois in Urbana, where he developed the fluorescence polarization technique for determination of membrane fluidity parameters. Since 1971, his research has been focused on various aspects of membrane structure and dynamics and currently his main interest is in introducing membrane fluidity methods to clinical diagnoses and treatments.

Dr. Shinitzky is an author of over 100 scientific publications and his research has been supported in part by grants and contracts from the National Cancer Institute.

CONTRIBUTORS

Yechezkel Barenholz, Ph.D.
Professor of Biochemistry
Department of Biochemistry and
  Neurochemistry
Institute of Biochemistry
Hebrew University
Hadassah Medical School
Jerusalem, Israel

J. G. Bluemink, Ph.D.
Hubrecht Laboratory
International Embryological Institute
Utrecht, The Netherlands

J. Boonstra, Ph.D.
Department of Molecular Cell Biology
State University of Utrecht
Utrecht, The Netherlands

Richard A. Cooper, M.D.
Professor of Medicine
Hospital of the University of
  Pennsylvania
Philadelphia, Pennsylvania

Andrew R. Cossins, Ph.D.
Department of Zoology
University of Liverpool
Liverpool, Merseyside, England

Siegfried W. de Laat, Ph.D.
Hubrecht Laboratory
International Embryological Institute
Utrecht, The Netherlands

R. Adron Harris, Ph.D.
Associate Professor
Denver V. A. Medical Center
University of Colorado School of
  Medicine
Denver, Colorado

Robert J. Hitzemann, Ph.D.
Associate Professor
Departments of Psychiatry and
  Pharmacology and Cell Biophysics
University of Cincinnati College of
  Medicine
Cincinnati, Ohio

Horace H. Loh, Ph.D.
Professor
Departments of Pharmacology and
  Psychiatry
University of California
San Francisco, California

Charles Everett Martin, Ph.D.
Associate Professor
Department of Biological Sciences
Rutgers University
New Brunswick, New Jersey

C. L. Mummery, Ph.D.
Hubrecht Laboratory
International Embryological Institute
Utrecht, The Netherlands

Gopa Rakhit, Ph.D.
Division of Drug Biology
Food and Drug Administration
Washington, D.C.

Michael Sinensky, Ph.D.
Associate Professor
Department of Biochemistry,
  Biophysics, and Genetics
University of Colorado Health Sciences
  Center
Eleanor Roosevelt Institute for Cancer
  Research
Denver, Colorado

Jerome F. Strauss, III, M.D., Ph.D.
Associate Professor
Department of Pathology and
  Laboratory Medicine
University of Pennsylvania
Philadelphia, Pennsylvania

Guy A. Thompson, Jr., Ph.D.
Professor of Botany
University of Texas
Austin, Texas

J. E. Thompson
Professor
Department of Biology
University of Waterloo
Waterloo, Ontario, Canada

W. J. van Blitterswijk, Ph.D.
Division of Cell Biology
The Netherlands Cancer Institute
Amsterdam, The Netherlands

Wieb van der Meer, Ph.D.
Biophysicist
The Netherlands Cancer Institute
Amsterdam, The Netherlands

P. T. van der Saag, Ph.D.
Hubrecht Laboratory
International Embryological Institute
Utrecht, The Netherlands

E. J. J. van Zoelen, Ph.D.
Hubrecht Laboratory
International Embryological Institute
Utrecht, The Netherlands

TABLE OF CONTENTS

Volume I

Volume II

Chapter 1

MEMBRANE FLUIDITY AND CELLULAR FUNCTIONS*

Meir Shinitzky

TABLE OF CONTENTS

*   Supported by Grant No. R01-CA-27471, awarded by the National Cancer Institute, Department of Health, Education and Welfare.

I. INTRODUCTION

It has long been realized that cell membranes have fundamental physiological tasks, in addition to their action as selective boundaries. Actually, most of the cellular biochemical and biophysical events occur in the membranes, where strict structural and dynamic features provide the control mechanisms. The effective two-dimensional structure of the membrane markedly reduces the number of degrees of freedom for protein-protein or protein-ligand interactions, which can increase by orders of magnitude the rates of processes as compared with an analogous isotropic system.[1-5] This is mostly facilitated by the directional diffusion, the increase in effective concentration and the reduction in number of possibilities for quaternary assemblage.[3,6]

Two principal mechanisms operate in membrane processes: one is mediated by metabolic energy and the other is directed by passive diffusion. Modulation of membrane activity is termed accordingly either as "active-modulation"[7,8] or "passive-modulation".[9,10] Membrane functional sites, which are anchored — chemically or physically — to the underlying network of microfilaments or microtubules, operate predominantly by metabolic energy. Their activity is characterized by inhibition with metabolic inhibitors or at low temperature and, in general, they mediate long-range effects in the cellular domain.[7,8] This mechanism is beyond the scope of this review, since the dynamics of the lipid layer play only a minor role there.

Passive-regulation of membranal function is determined by the protein-lipid interplay, where the lipid dynamics play the dominant role. By and large, the lipid structure — under physiological conditions — can be envisaged as a two-dimensional array where diffusion processes, though complex,[11] can be resolved into in-plane and out-of-plane modes. The translational diffusion can be resolved into "lateral" (in-plane) or "vertical" (out-of-plane) motions, whereas the rotational diffusion can be resolved into "uniaxial" (in-plane) and "flip-flop" (out-of-plane) rotations. The real diffusion pattern is probably a combination of the in-plane and out-of-plane modes. Through these modes, a diffusion and displacement passive-regulation operates and, unlike the active-regulation, it is independent of metabolic energy and, in principle, can take place in isolated membranes.

This chapter presents an overview on the current state-of-the-art of membrane fluidity. Comprehensive accounts on each of the mechanisms involved in alteration of the membrane fluidity, as briefly reviewed in this chapter, are presented in the following chapters.

II. THE LIPID FLUIDITY OF CELL MEMBRANES

Under physiological conditions the lipid constituents of biological membranes diffuse almost freely in the plane of the membrane. In the hydrocarbon core of the lipid layer, the dynamic features resemble a macroscopic hydrocarbon fluid, although with a series of fundamental differences. Basically, the lipids in membranes are arranged in the form of a bilayer, with two principal directions of diffusion — along and across the plane of the membrane. The diffusion processes, confined to this geometry, are therefore complex and attempts to resolve it on grounds of two-dimensional diffusion were only partially successful.[12,13] Other features which add to this complexity are the variation of dielectric constant[14,15] and the asymmetric distribution of phospholipids in the endofacial and the exofacial sides of the membrane.[16,17] Asymmetry can also be present in the same layer, where lipids can form domains of different composition and diffusion characteristics.

In an attempt to circumvent this complexity, the fluidity of lipids can be operationally confined to the hydrocarbon region where the asymmetric features are less pronounced than those around the lipid headgroups. The operational term used in this approach is microviscosity ($\bar{\eta}$), which is expressed in macroscopic units (reviewed in References 18 and 19). The lipid microviscosity actually simulates the anisotropic lipid core with an equivalent isotropic fluid and is, therefore, of low microscopic resolution and can be viewed as a submacroscopic scale of lipid viscosity (see below). This simulation can be extended to classical hydrodynamic expressions, which can be applied to the membrane lipid domain. The most important are the exponential expression, which describes the dependence of $\bar{\eta}$ on temperature (Equation 3), and the stationary expression, which describes the inverse relation between $\bar{\eta}$ and the free volume (Equation 4).

From the teleological point of view, the integrity of the submacroscopic lipid microviscosity is reflected in processes of adaptation to temperature or nutritional stress, where the membrane lipid microviscosity is temporarily impaired. Shortly after imposition, a series of lipid biosyntheses for restoration of the membrane fluidity are triggered. The ensuing changes in membrane lipid structure and composition are almost entirely confined to the hydrocarbon layer, without any substantial change in the lipid headgroup region. It seems, therefore, that the lipid hydrocarbon layer is the most sensitive region to physical and chemical effects. Maintenance of the lipid microviscosity at a precise level is presumably a prerequisite for a proper physiological functioning. Despite its shortcomings, the term "lipid microviscosity" is still the most practical among the other alternative physical terms which could be applied to lipid dynamics. Only through this term it is now possible to offer semiquantitative assessment on the dynamic effects of lipids on membranal functions.

A. The Submacroscopic View

The dynamic details which together determine the "lipid fluidity" can, in principle, be resolved by analyzing the movement of each atom or segment in the lipid chains, with the aid of proton or C nuclear magnetic resonance.[20,21] The resulting microscopic view of lipid fluidity is of a complex combination of position, orientation, order, and motional freedom of the various hydrocarbon moieties.[20,21] In other methods, like electron spin resonance (ESR)[22] and fluorescence depolarization,[18] a foreign probe is introduced which, by its spontaneous distribution and movement, averages out most of the microscopic details and therefore reports on the submacroscopic level of lipid fluidity.[19,23]

At the submacroscopic level, the "true" lipid microviscosity, $\bar{\eta}$, is a tensorial parameter which can be presented as[23]

$$\bar{\eta} = \frac{1}{3}\,\eta_{\parallel} + \frac{2}{3}\,\eta_{\#} \tag{1}$$

Table 1
EVALUATION OF LIPID MICROVISCOSITY BY STEADY-STATE FLUORESCENCE POLARIZATION

Level of $\bar{\eta}$ presentation	Fluorescence polarization relation	Comments	Ref.
Qualitative	$\left(\dfrac{r_0}{r} - 1\right)^{-1}$	Highly nonlinear	18, 26
Quantitative on a relative scale	$2.4\left(\dfrac{r_0}{r} - 1\right)^{-1}$	Approximate; useful in measurements of the same system at different conditions	18
Approximate quantitative for DPH (in poise)	$C \cdot T \cdot \tau\left(\dfrac{r_0}{r} - 1\right)^{-1}$	Apparent value	18
Apparent quantitative (in poise)		For freely rotating probes; a combination of the true viscosity and the degree of order	18, 25, 26
"True" quantitative for DPH	$\bar{\eta}_{apparent} \cdot \left(\dfrac{0.089}{r} - 0.116\right)$ for $0.13 < r < 0.28$	Defined as $1/3\eta_\| + 2/3\eta_{\#}$, considerably smaller than the apparent $\bar{\eta}$	23, 31

where $\eta_\|$ and $\eta_{\#}$ are the principal viscosity vectors across and along the plane of the lipid bilayer. Partial resolution of $\eta_\|$ and $\eta_{\#}$ can be obtained with a disc-like fluorescence probe (e.g., perylene), by determination of the apparent viscosities which oppose its in-plane and out-of plane rotations.[24-26] Alternatively, and more directly, measurements of lateral diffusion coefficients of fluorescent[27] or ESR probes[28] can be correlated predominantly with $\eta_{\#}$. The difference between $\eta_\|$ and $\eta_{\#}$ is reflected in a residual fluorescence anisotropy, r_∞, of a rod-like probe (e.g., DPH), after a pulse excitation.[29,30] In a fluid system $\eta_{\#} \sim \eta_\|$ and $r_\infty \to 0$ and in rigid system $\eta_{\#} \gg \eta_\|$ and $r_\infty \to r$.[31]

Since most $\bar{\eta}$ values in the literature were derived by steady-state fluorescence depolarization of DPH,[18] without corrections for r_∞, they are actually apparent values which combine both the true $\bar{\eta}$ value with the packing anisotropy of the system, as reflected in r_∞. The approximate "true" $\bar{\eta}$ can be derived, however, by multiplying the apparent $\bar{\eta}$ value — estimated from steady-state fluorescence polarization — by the correction factor $(1 - [r_\infty/r])/(1 - [r_\infty/r_0])$, where r and r_0 are the measured and the limiting fluorescence anisotropy values.[23,29] Based on recent data,[31] this correction factor was estimated to be equal to $(0.089/r) - 0.116$.[23] A brief summary of the various levels of evaluation of $\bar{\eta}$, by means of steady-state fluorescence polarization, is given in Table 1. The term r_∞/r_0 is the square of the order parameter in the system.[29,31] Therefore, the more rigid is the bilayer, the contribution of order to the apparent $\bar{\eta}$ value increases due to probe alignment.

For most physiological functions related to lipid fluidity, the submacroscopic parameters are relevant, though it is not yet clear whether the true or the apparent $\bar{\eta}$ is of greater relevance. This is mainly reflected in the fact that modulation of a membrane function by alterations of lipid composition (e.g., cholesterol level or degree of unsaturation of the phospholipid acyl chains) corresponds to similar changes in microviscosity, irrespective of the type of lipid change, and with only little dependence on its specific microscopic details. Also, in vivo restoration of an impaired membrane microviscosity proceeds mostly by changes in either cholesterol level or the composition of phospholipid acyl chains, with the possibility of mutual interchange. The use of the submacroscopic term "lipid microviscosity" is therefore of direct relevance to membrane function, and in vitro or in vivo manipulations of membrane microviscosity by natural lipids are expected to modulate the membrane functions to the same degree. The various factors which determine the membrane microviscosity are discussed in the following. Detailed mechanisms relating to alterations of these factors are described in the other chapters of this book.

B. Chemical Modulators

The chemical composition of membrane lipids can change either by translocation or exchange processes with the exterior, or when induced by syntheses either *in situ* or intracellularly. In general, modulation of lipid composition from the onset signal to the new steady-state takes minutes or hours, depending on the type of process. Therefore, apparent changes of lipid fluidity at shorter time intervals cannot be accounted for by chemical modulation.

The main natural modulators of membrane lipid fluidity are presented below in their presumed order of significance. Each of these modulators induces considerably different microscopic diversity in the lipid domain; yet, the submacroscopic approach (Section II.A) contends that these are of only secondary importance. In cases where a specific chemical modulator can also react with a certain functional site, its submacroscopic effect on the lipid fluidity — when this particular site is being studied — is of only partial relevance.

1. Cholesterol

Under physiological conditions, cholesterol acts as the main lipid rigidifier in natural membranes.[32,33] Its effect is mediated both by overall increase in $\bar{\eta}$ and by increase in order of the lipid bilayer.[31,32] The latter effect is displayed by either reduction in ΔE (see Equation 3) or increase in r_∞ (see Section II.A). It should be noted, however, that below the lipid phase transition, where a pure lipid bilayer is highly ordered, cholesterol acts in an opposite way — it induces lipid melting with the overt increase in fluidity and decrease in order.[32,34] Lipid domains below the phase transition, however, are very uncommon at the physiological state.

The molecular features of the structural integration of cholesterol in lipid bilayers[35-37] imply that it may have only insignificant preference towards certain phospholipids. Yet, sphingomyelin and other rigidifying phospholipids were shown to associate with cholesterol better than the other phospholipids.[38,39] In a good approximation it can still be assumed that, up to a certain level, cholesterol is distributed evenly between the phospholipids, and the mole index of cholesterol/phospholipids (C/PL) can serve as a good qualitative parameter for correlation with the submacroscopic lipid microviscosity.[40,41] At abnormally high cholesterol levels (C/PL>2) a sharp transition in cholesterol organization, presumably to segregated domains, takes place, such that further addition of cholesterol does not increase the membrane microviscosity.[42,43]

The weak and nonspecific association of cholesterol with phospholipids resembles a solute-solvent type system. Thus, cholesterol can exchange freely between various phospholipid pools,[43-56] with a partitioning directed from cholesterol-rich to cholesterol-poor reservoirs. This process of cholesterol translocation is largely determined by the C/PL of the participating pools and is especially pertinent to cholesterol exchange between the membranes of blood cells and the serum lipoproteins.[42,48,49,57,58] Another process, which can modulate the level of membrane cholesterol, involves specific interaction of serum lipoproteins with surface receptors, followed by internalization and disintegration.[59]

Metabolic changes in cholesterol level can also operate, on the one hand, via intracellular degradation or synthesis and, on the other hand, by the formation or hydrolysis of cholesterol esters. The latter family of compounds is abundant in all tissues, especially in the serum, but — unlike unesterified cholesterol — it is excluded from the lipid bilayers of biological membranes. One of the most efficient cellular mechanisms for elimination of excessive membrane cholesterol operates through esterification and the exclusion of the cholesterol esters formed.[60,61] Similarly, esterification and de-esterification of serum cholesterol provide an efficient preservation of a constant C/PL, which in turn maintains a constant level of membrane fluidity.[62]

2. Degree of Unsaturation

Introduction of a *cis*-double bond into a phospholipid acyl chain induces a marked increase in specific volume which is expressed in reduction of viscosity (see Equation 4). The double bond has the greatest effect when it is introduced into a fully saturated chain (for example, when a stearoyl (18:0) chain is converted to an oleoyl (18:1 *cis*) chain). The introduction of a second double bond, as from oleoyl to linoleoyl, has a much lesser effect on reduction of viscosity. At higher degrees of unsaturation, the effect becomes progressively less pronounced.[63-67]

The degree of unsaturation of phospholipid acyl chain is determined, to a major extent, at the level of membrane biogenesis, where the proper fatty acids are being selected in the process of phospholipid biosynthesis. Special intracellular enzymes ("desaturases") can manipulate the number of double bonds[68] and the length[69] in the available pool of fatty acids. Exchange processes, either spontaneous or with the aid of special protein carriers,[70-73] can also induce a net change in the overall degree of unsaturation, but these processes are presumably less effective than the membrane biogenesis pathway.

3. Sphingomyelin

The two phosphorylcholine phospholipids, lecithin and sphingomyelin, comprise over 50% of the phospholipid contents in both biological membranes and fluids. Except for the common headgroup, these two phospholipids are highly dissimilar in physicochemical properties (for a recent review, see Reference 74). The gel-liquid crystalline phase transition of natural sphingomyelin appears around 37°C, whereas for natural lecithin it is at a much lower temperature. Therefore, around the physiological temperature, there is a difference between the lipid microviscosity in bilayers comprised of either one of these phospholipids.[24] In fact, natural sphingomyelin and lecithin are at the extreme edges of contribution to rigidification (sphingomyelin) or fluidization (lecithin). Besides imparting rigidity on lipid layers, sphingomyelin is a powerful coupler of the two lipid monolayers;[75] it can form separate domains[76] and it seems to have affinity to cholesterol[39] and proteins.[77]

It is of interest that in various processes, where the relative content of lecithin or sphingomyelin is altered, their total amount remains almost constant.[74] This implies that these phospholipids are somehow interchangeable. Therefore, changes in lipid fluidity — due to changes in sphingomyelin content — can be presented in terms of changes in lecithin-to-sphingomyelin mole ratio, L/S.[78,79] This ratio was found to be the major factor which contributes to lipid fluidity modulation in abetalipoproteinemia,[80] in muscular dystrophy,[81,82] in the lung surfactant of newborns,[79,83] and in aging of tissues.[84-88]

4. Phosphatidylethanolamine

The primary amine headgroup of phosphatidylethanolamine has the ability to form hydrogen bonds with an adjacent phosphate group. Since all other structural features of phosphatidylethanolamine are virtually identical to those of phosphatidylcholine (lecithin), it is expected that the viscosity imparted by phosphatidylethanolamine is higher than that of lecithin. This assumption has been verified in a series of studies by Hirata and Axelrod (reviewed in Reference 89), who discovered two methylating enzymes that reside in biological membranes and can sequentially methylate phosphatidylethanolamine to phosphatidylcholine with the net increase in lipid fluidity. This conversion also transverses phosphatidylethanolamine from the inner monolayer of the membrane to the outer monolayer, which is the natural residence of lecithin.[16]

Methylation of phosphatidylethanolamine is a subtle mechanism for lipid fluidization and probably takes part in signal transduction induced by hormones, neurotransmitters, and mitogens.[89]

5. *Other Membrane Lipids*

The remaining common phospholipids, i.e., phosphatidylserine, phosphatidylglycerol, and phosphatidylinositol, are all of a relatively high degree of unsaturation and may, therefore, be considered as fluidizers, similar to lecithin.[19] The sphingoglycolipids (e.g., gangliosides), on the other hand, have a hydrophobic region which is similar to that of sphingomyelin, and as such they act as lipid rigidifers.[90,91] The effect on fluidity of other lipids, which occasionally appear in biological membrane — in particular, etheric phospholipids (i.e., plasmalogens) — remains to be studied.

6. *The Protein Content*

In general, in contrast to the lipid bilayer, proteins are of very low compressibility.[92,93] Occupying a volume of about 50 to 100 phospholipids, membrane proteins can be envisaged as bulky rigid domains in the lipid fluid matrix. The thermal motion of lipids in the vicinity of the proteins is expected to be hindered markedly with a local increase in microviscosity. This effect fades as the distance from the protein increases, and the net effect on the overt lipid microviscosity should therefore increase with the increase of the protein content. For most membranes, where the protein content is approximately 50% by weight, the local effect of each individual protein remains pertinent only to the boundary layer of lipids, the ‘‘lipid annulus’’,[94] which is highly immobile,[95,96] while at more distal lipid domains the effects of the other proteins converge to a more or less homogeneous rigidification. The lipid rigidification effect of proteins was verified experimentally and could be best demonstrated by comparing the membrane lipid microviscosity to that of liposomes prepared from the membrane total lipid extract.[33,40,97,98]

In essence, the effect of proteins on lipid dynamics is similar to that of cholesterol.[33] Both rigidify and increase the order in fluid lipid domains and act conversely below the lipid phase transition. It is also interesting that the mutual effects of cholesterol and protein apparently maintain the flow activation energy (Section II.C.1) of the membrane lipid layer within a narrow range which is presumably a prerequisite for proper physiological functioning.[19,33]

C. Physical Effectors

The membrane lipid fluidity is affected by the ambient physical factors. The relaxation of the membrane microviscosity, after a physical perturbation, is expected to be completed within less than 1 sec[99,100] which, on a physiological time scale, may be considered instantaneous.

In general, a change in a thermodynamic parameter (e.g., temperature or pressure) can affect both the steady-state organization of the lipids and the dynamics of the flow processes. This can be expressed in a general Boltzmann-type distribution function:

$$\overline{\eta}_x = \eta_{ox}\, e^{\Delta E_x/RT} \tag{2}$$

where x is the modulated parameter, $\overline{\eta}_x$ is the apparent microviscosity at a particular x, $\Delta E x$ is the specific activation-energy of flow which relates to x, and η_{ox} is the limiting steady-state value of $\overline{\eta}_x$ which can be reached when $\Delta E x \rightarrow 0$. For most cases, e.g., upon temperature changes in the physiological relevant range of 0 to 40°C (i.e., 273 to 313 K), η_{ox} is close to a constant and, therefore, $\overline{\eta}_x$ should vary with x in a simple exponential manner.

1. *Temperature*

One of the most important expressions extrapolated by the macroscopic simulation of lipid fluidity is the exponential dependence of the microviscosity ($\overline{\eta}$) on the reciprocal of the absolute temperature (T):

$$\overline{\eta} = A\ e^{\Delta E/RT} \qquad\qquad (3)$$

where ΔE is the thermal activation energy.[25,33] At physiologically relevant temperatures, where the lipid layer is maintained in its liquid-crystalline phase, a plot of log $\overline{\eta}$ vs. 1/T should yield a straight line — a criterion which is well obeyed in most biological membranes.[18,33] Lipid phase-transitions or phase-separations are displayed by abrupt breakpoints or curvatures in such plots.[18,65]

The flow activation energy, ΔE, characterizes the sensitivity of $\overline{\eta}$ to changes in temperature, and in addition provides a criterion for the degree of order in the lipid core.[33] Most of the biological membranes in mammalian systems possess a ΔE value around 7 kcal/mol,[33] which corresponds to an increase in $\overline{\eta}$ by a factor of 1.6 or 3.8 when the temperature is decreased from 37°C to 25°C or 4°C, respectively. Plant membranes are characterized by a much lower ΔE value of around 3.5 kcal/mol,[101] and the increase in $\overline{\eta}$ upon the same decrease in temperature is only by a factor of 1.2 and 1.9, respectively. The low ΔE in plant membranes may reflect a highly ordered hydrocarbon core, which is only mildly affected by changes in the ambient temperature. This is probably a teleological prerequisite for a proper resistance to temperature change in plants.

Temperature is the most critical energetic parameter in virtually all biological systems. Besides lipid fluidity, it affects simultaneously and indiscriminately the function, stability, and vitality of all microscopic entities in each cell. In plants, microorganisms, and poikilotherms, temperature changes are of physiological relevance and it can be assumed that they are well-equipped with specific tools for regulation of membrane function, in response to changes in temperature (see Section II.D.4).

2. Volume

Upon swelling, shrinkage, or mechanical deformation of organelles, the average distance between the lipid constituents in the membrane can change. The changes ensued in lipid specific volume should correlate with changes in $\overline{\eta}$. In a series of aliphatic hydrocarbons and other organic fluids, where the main cohesive forces are van der Waal's attractions, this correlation is given by the empirical equation:[23,102,103]

$$\eta = \frac{B}{V - V_\infty} = \frac{B}{\Delta V} \qquad\qquad (4)$$

where η is the viscosity, V is the specific volume (i.e., the reciprocal of density), V_∞ is the limiting value of V (at infinite viscosity), and B is a proportionality factor which increases with the reduction in degrees of freedom in the series. The term ΔV is the free volume of the fluid and, as expressed in Equation 4, is proportional to the fluidity (ϕ) and inversely proportional to the viscosity. Combination with Equation 3 yields:

$$\Delta V = \Delta V_0\ e^{-\Delta E_v/RT} \qquad\qquad (5)$$

where $\Delta E_v \sim - \Delta E$, which could be verified experimentally.[23] This reciprocity between density and fluidity actually prevails in most fluids.[51,104-107]

Equation 4 holds exceptionally well for nonassociated organic fluids, like aliphatic hydrocarbons and their derivatives.[23,108] In principle, the submacroscopic approach permits the application of Equation 4 to lipid bilayers, since both their packing and thermotropic behavior are largely determined by the hydrocarbon region. Verification of this important feature is shown in Figure 1, where volume-area parameters — taken from compressibility experiments of lipid monolayers[109] — were compared with the lipid microviscosity of lipid bilayers of identical composition.[23] As clearly demonstrated, the lipid microviscosity is inversely cor-

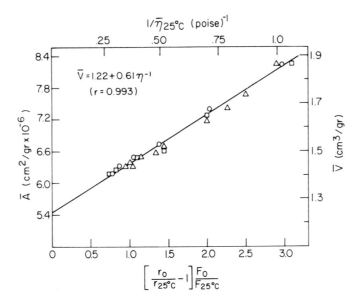

FIGURE 1.   Correlation between lipid fluidity ($1/\bar{\eta}$) and specific surface area, $\bar{A}$, or volume, $\bar{V}$, in mixtures of egg lecithin and cholesterol at various mole ratios. Determination of $\bar{\eta}$ was carried out in sonicated liposomes, using DPH as a fluidity probe. Data for lipid specific area were taken from experiments in monolayers[109] and were normalized to the average molecular weight. The specific volume was calculated for a thickness of $(CH_2)_{18}$, taking increments of 1.26 Å per CH_2. (From Shinitzky, M. and Yuli, I., *Chem. Phys. Lipids*, 30, 261, 1982. With permission.)

related to the free volume (or free area) of the lipid constituents. The B value of this correlation is much higher than that of aliphatic hydrocarbons (0.61 erg·sec for the lipid bilayer, as compared to $7.5 \cdot 10^{-4}$ erg·sec for the hydrocarbon liquids, Reference 24), which indicates a marked decrease in the number of degrees of freedom for flow presumably due to the headgroup interactions. The applicability of Equation 4 to lipid bilayers is further verified by the invariable V_∞ value: 1.22 cm³/g for lipid bilayer (Figure 1) and 1.28 cm³/g for aliphatic hydrocarbons.[102,108,110]

Under constant physical conditions, B remains approximately constant and changes in $\bar{\eta}$, due to change in composition, correspond almost exclusively to the changes in ΔV. Since the latter is a basic thermodynamic parameter, one can use this important interrelation to bridge the operational parameter, $\bar{\eta}$, and the thermodynamic state of the system. Furthermore, changes in $\bar{\eta}$ by one of the chemical modulators (Section II.B) can be related to changes in ΔV. In other words, a lipid rigidifier (e.g., cholesterol) can be assumed to render the system more viscous almost exclusively by reducing ΔV (condensing effect). If one further assumes that the change in $\bar{\eta}$, induced by one of the modulators, does not affect the thickness of the bilayer, then between two states the following relation should hold:

$$\bar{\eta}_1/\bar{\eta}_2 = (A_2 - A_\infty)/(A_1 - A_\infty) \qquad (6)$$

Thus, the relative change in volume or surface area (A) is smaller in magnitude than the associated change in $\bar{\eta}$. Alternatively, the change in fluidity corresponds to

$$\phi_2 - \phi_1 \propto A_2 - A_1 \quad \text{or} \quad \Delta\phi \propto \Delta A \qquad (7)$$

3. Pressure

The energetic parameter $P\Delta V$ implies that yielding to pressure (P) corresponds to a change in free volume (ΔV). In as much as water and proteins are considerably less compressible than the lipid bilayer, pressure effects on biological membranes correlate almost exclusively with changes in lipid free volume (see before). This is primarily due to changes in lipid surface area and is therefore manifested in changes in $\overline{\eta}$ or the degree of order.[111]

The effects of pressure on the lipid microviscosity are, in principle, different for scalar (homogeneous) and vectorial pressures. Under homogeneous (e.g., hydrostatic) pressure it can be assumed that the main effect on lipid microviscosity is mediated by the addition of a pressure-volume energy barrier in the flow activation:[106]

$$\overline{\eta}_p = \overline{\eta}_0 \, e^{P\Delta V^{\neq}/RT} \tag{8}$$

$$\phi_p = \phi_0 \, e^{-P\Delta V^{\neq}/RT} \tag{9}$$

where $\overline{\eta}_p$, $\overline{\eta}_0$, ϕ_p, and ϕ_0 are the microviscosities or fluidities at P and zero pressure, and $\Delta V^{\neq}$ is the change in molar volume during the flow activation. In the more general case, the viscosity of fluidity can reach limited values (η_∞ or ϕ_∞) at a certain P, above which they do not change any further. Equation 9 then becomes:

$$\phi_p = (\phi_0 - \phi_\infty) \, e^{-P\Delta V^{\neq}/RT} + \phi_\infty \tag{10}$$

At ϕ the compressibility of the lipid layer presumably approaches zero, but the thermal energy maintains a finite fluidity which can be decreased by reducing the temperature. As an example, the change in lipid fluidity with hydrostatic pressure is shown for egg lecithin liposomes in Figure 2. Verification of Equation 10 is presented as

$$\log (\phi_p - \phi_\infty) = C_1 P \tag{11}$$

in the inset (C_1 is a constant).

Phase transition of lipid assemblies is characterized by a high ΔE. Therefore, relatively small changes in either temperature or pressure around the transition point can shift the lipid phase.[112] The interrelation between temperature and pressure, dT/dP, at the phase transition is in the range of about 20°C/1000 atm for both lipid dispersions[113-115] and biological membrane.[116,117] A net change in the lipid free volume of about 3 to 4% was estimated to take place in the melting process.

Osmotic pressure, as a physiologically relevant vectorial pressure, can cause swelling or shrinkage of organelles and, thereby, stretching or condensing of their membranes, which in turn (Section II.C.2) can change the lipid microviscosity[118] or the lipid phase.[113] In a spherical enclosure (e.g., a bubble, liposome, or spheroid erythrocyte) of a radius r, a pressure-volume equilibrium persists which obeys the Young-Laplace equation[119]

$$P = 2\gamma(r)/r \tag{12}$$

where P is the excess internal pressure and $\gamma(r)$ is the tension constant of the surrounding membrane, which resists surface expansion and is dependent on r. In liposomes or bubbles, $\gamma(r)$ is the surface tension, while in membranes it represents a complex viscoelastic constant. A system with internal pressure, like a detached bubble, is stable only if γ increases sufficiently with r (or surface area). At a constant γ, the volume will spontaneously increase, so as to decrease P until rupture. In the case of an incomplete bubble, which protrudes from a planar surface, r remains approximately constant since the change in surface area arises

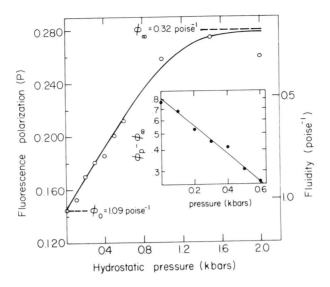

FIGURE 2. Increase in the degree of fluorescence polarization of DPH labeled egg-lecithin liposomes upon application of hydrostatic pressure and the corresponding decrease in fluidity. The data are presented according to Equation 11 in the inset. (Communicated by P. Chong.)

from material which is supplemented from the surrounding plane, rather than from surface stretching.

For evaluating the relation between osmotic pressure, π, and lipid microviscosity in spherical organelles, one can use the following form of Equation 12:

$$\pi = C_2 \frac{\gamma(A)}{\sqrt{A}} \tag{13}$$

where A is the surface area. A good representation of $\gamma(A)$ could be a linear function[119-122] where γ remains constant (γ_0) as long as no change in the resting radius, r_0 takes place, while upon stretching, r increases by increments proportional to the added radius,

$$\gamma(A) = \gamma_0 + C_3 \, \Delta r/r_0 \tag{14}$$

A similar linear relation presumably applies to condensation. Assuming that this simple relation indeed holds, the dependence of the lipid fluidity ϕ on the osmotic pressure will obey the following approximate relation (see Equation 7):

$$\sqrt{\Delta\phi} = C_4 \, (\pi - \pi_0) \tag{15}$$

where π_0 is the limiting osmotic pressure at the resting surface area A_0. Verification of Equation 15, in sealed phospholipid vesicles, is shown in Figure 3.

Pressure can affect virtually all physiological functions.[123] Since membrane lipid layers are by far more compressible than proteins, it is reasonable to assume that the observed pressure effects on cellular activities[123-127] originate predominately from the corresponding changes in the membrane microviscosity. In this respect, one of the most investigated pressure-function relations has been the reversal by pressure of the effect of drugs and narcotics (e.g., alcohol) which induce lipid fluidization.[123,128-132] A physiologically relevant

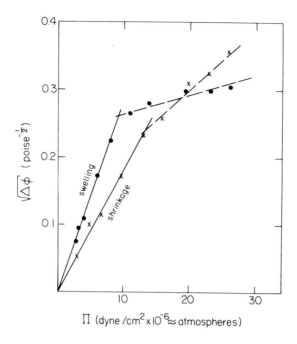

FIGURE 3. Change in lipid fluidity with osmotic pressure, presented according to Equation 15, in sealed unilamellar liposomes composed of 96% egg lecithin and 4% dicetyl phosphate. Liposomes were prepared in phosphate-buffered saline (PBS) for shrinkage experiments, or in PBS + 1 M sucrose, for swelling experiments, and then diluted into PBS containing 0—1 M sucrose. Net osmotic pressure, π, was estimated from a correction table and lipid fluidity, ϕ, was assessed by DPH fluorescence depolarization.

pressure-microviscosity interplay takes place in the lung surfactant during respiration. It has been proposed that the special composition of the lung surfactant is characterized by a $\gamma(r)$ (see Equation 14) which, around the body temperature (37°C), where the lipid organization undergoes a quasi phase transition, remains constant.[119]

4. Membrane Potential

Formation of electric potential across a lipid bilayer causes a small but significant increase in $\bar{\eta}$, both in artificial vesicles[133,134] and in natural membranes.[135] The reverse correspondence, namely decrease in membrane potential upon increase in lipid fluidity, was also observed.[136] Since structural deformation is presumably too small to account for this effect,[133] it could be correlated with the imposition of electrical energy barrier in the flow activation, in addition to the prevailing thermal energy barrier ΔE (see Equation 3). Similar to the effect of pressure (Section II.C.2), the membrane microviscosity at a potential ϵ can be presented by:

$$\bar{\eta}_\epsilon = \eta_{\epsilon=0} \; e^{\epsilon \Delta Q^{\neq}/RT} \qquad (16)$$

where $\Delta Q^{\neq}$ is the transient difference of charge distribution between the activated state and the resting states of flow. According to Equation 16, the effect of membrane potential on $\bar{\eta}$ should mostly depend on the magnitude of $\Delta Q^{\neq}$. In artificial vesicles, $\Delta Q^{\neq} \simeq 3 \times 10^3$ C/mol[133] or ~3% ionic equivalents, which corresponds to about 10% increase in $\bar{\eta}$ upon application of a membrane potential of 100 mV. Transient changes in membrane potential of this magnitude take place in various electrophysiological processes and may provide a subtle means for transitory modulation of the diffusion, as well as the degree of exposure, of membrane proteins (Sections III.A and III.B).[137-139] In addition, the mutual dependence of membrane potential and osmotic pressure[125-127] could be mediated via the corresponding change in lipid fluidity.

5. Acidity

Hydrogen bonds and electrostatic interactions at the polar headgroup region of the lipid bilayer are responsible for the increase of about three orders of magnitude in the viscosity of the hydrocarbon core, in comparison to analogous aliphatic hydrocarbon (Section II.C.2). Since the free volume characteristics of the lipid core and an equivalent hydrocarbon fluid are similar, the headgroup interactions presumably mediate their effect both by increase of the energy of association and decrease of the rate of shuttling. These are expressed in increase of B value in Equation 4 or increase of A value in Equation 3. As a result, changes in charge distribution at the polar headgroups by pH changes can induce a change in lipid microviscosity.[140-143] The phospholipids, which are most susceptible to changes in pH around pH = 7, are phosphatidylserine with a carboxyl of a high pK and phosphatidylethanolamine with an amine of a low pK. The abnormal pKs of these side chains can be mediated directly or indirectly by neighboring protein side chains.

6. Calcium

In general, nonspecific effects of Ca^{+2} are based on crosslinking of negative charges. Except for lecithin, sphingomyelin, and phosphatidylethanolamine, all other natural phospholipids are negatively charged and as such are susceptible to crosslinking by Ca^{+2}. At low level of interaction, Ca^{+2} may induce some structural rearrangement of the phospholipids with only a small net effect on the overt microviscosity. Stronger interactions, however, may cause a significant rearrangement which, in the extreme cases, may lead to phase separation of negatively charged lipids.[140,144] Ca^{+2} may also provide bridging between proteins and lipids, which could exert a marked effect on the membrane assembly.[145]

D. Physiological Determinants

1. The Cell Cycle

A series of membrane activities change during the cell cycle[146,147] some of which — like transport, enzymic activities, and receptor expressions — can be related to changes in the membrane lipid fluidity. As discussed in Section III, a substantial portion of the observed changes during the cell cycle is associated with passive modulation of membrane components, rather than the biosynthetic cycling of the plasma membrane.

The data reported in the literature do not indicate any general profile for the change in membrane lipid fluidity or protein mobility during the cell cycle. Neuroblastoma cells in mitosis acquire higher membrane microviscosity[148,149] or lower protein mobility[149] than in the other phases. In the S phase these properties reverse, while during the G_1 and G_2 interphases the corresponding changes in lipid fluidity or in protein mobility take place.[148,149] A converse profile of lipid fluidity, namely a lower lipid microviscosity in mitosis than in the S phase or the interphases, was observed in lymphocytes,[150] fibroblasts,[150] hepatocytes,[150] hepatoma cells,[44] and Chinese hamster ovary cells.[151] No change in membrane lipid fluidity was detected in the cycle of mouse leukemic cells.[152] It seems that the change in membrane lipid fluidity during the cell cycle is programmed to adjust the specialized functions of each phase. Part of these changes could be associated with changes in water permeability and osmotic pressure.[128]

2. Differentiation and Maturation

This stage of cellular differentiation can be characterized by maturation processes which are progressively slowed down. Most cellular processes reach basal activity, concomitantly with an increase in membrane microviscosity (i.e., cholesterol, References 153 to 157), and only specific effectors can stimulate these processes to levels which are found in undifferentiated cells. Similarly, differentiation is affected by a progressive increase in microviscosity of the cell membrane, as has been documented for Friend erythroleukemia cells,[158] promyelocytic leukemia cells,[159] and neuroblastoma cells,[160] which is associated, at least in part, with an increase in cholesterol level[153,158] (see Section II.B.1).

It is not yet clear whether the increase in the cell membrane microviscosity associated with differentiation is due entirely to intracellular processes of membrane biogenesis, or whether serum lipids (e.g., cholesterol) also contribute to it. If the latter process takes place in differentiation, manipulation of serum lipids could in turn modulate the rates of cellular differentiation and maturation. In line with this possibility, it has been observed that the rate of cell proliferation is inversely proportional to the membrane microviscosity or the cholesterol level.[47]

3. The Cell Density

Towards confluency, or at the stationary phase of cell growth, the rate of proliferation is markedly reduced concomitantly with an increase in membrane microviscosity.[47,161-163] As in differentiation, this increase is most closely related to increase in cholesterol level.[47] When cells in confluency are diluted abruptly, they acquire the expected reduced membrane microviscosity within a relatively short period of time.[163] This indicates that the cell-cell encounters determine, to a major extent, the increase in membrane microviscosity and the related rate of proliferation.[163]

The apparent correlation between "contact inhibition" of cell proliferation and increased membrane microviscosity raises the intriguing possibility that the abnormally low membrane microviscosity of tumor cells[164] may be part of the cause of their unrestrained proliferation.[165]

4. Aging

Two phases can be ascribed to processes of tissues aging both in vivo and in vitro. In the early phase of aging, the rate of metabolic pathways is slowed down to such a level that it cannot fully cope with external stress. At this phase, when lipids from the serum (e.g., cholesterol) partition into the cell membrane, the rates of the homeostasis adjustments (Section II.D.5) are presumably slower than the rate of lipid accumulation, with the net progressive increase in membrane microviscosity. A similar relation probably develops between intracellular membranes and the cytosol. The early phase of aging is, therefore, characterized by a change in composition of membrane lipids, with a net increase in microviscosity. In principle, aging at this phase can be reversed by lipid manipulation (see Section IV).

At the advanced state of aging the protein population in the membranes is altered as a result of shedding, enzymic degradation, oxidation, crosslinking, or genome aberration. Unlike the early phase, the advanced phase is, to a major extent, irreversible.

Table 2 summarizes the changes observed in lipid composition of membranes and tissues with aging. Essentially, all reported changes indicate an increase in microviscosity. Changes in lipid composition with age of nerve tissues were summarized in a review by Rouser et al.[86]

5. Adaptation

In a healthy tissue the lipid fluidity of the cell membranes is at a well-defined state, which maintains optimal function. States of stress, which are induced by continuous perturbation

Table 2
CHANGES IN LIPID COMPOSITION WITH AGING

The increased lipid (relatively to the others)

Tissue	Animal	Cholesterol	Sphingomyelin	Glycosphingolipids	Degree of saturation	Ref.
Brain	Human	+	+	+		86, 87
Brain	Rat	+	+	+		88
Myelin	Rat	+	+			153
Myelin	Rat	+	+	+		154
Aorta	Human	+	+	+	+	85,166,167
Aorta	Rat		+	+		168
Adipocyte membrane	Rat				+	155
Lymphocyte	Human	+				169
Membrane	Mouse	+				170
Erythrocyte membrane	Human	+				171

of membrane lipid fluidity, generally mediate a homeostatic response for restoration of the lipid fluidity to its optimal state.[172,173] This process, which was termed "homeoviscous adaptation",[174] proceeds mainly via alteration of the degree of unsaturation of the phospholipid acyl chains[175-179] or changes in the level of cholesterol.[179-182] It is interesting that under stress conditions the hydrocarbon layer is the only region which is affected significantly and ultimately restored. Furthermore, the microscopic differences between lipid systems, maintained at constant fluidity by changes in either the degree of unsaturation of the phospholipid acyl chains or the level of cholesterol, seem to be of only secondary importance. This is reflected in mutants defective in cholesterol biosynthesis[178] or in cells depleted of cholesterol,[196,231] where adaptive compensation proceeds via decrease in the degree of unsaturation of the phospholipid acyl chains. Since the polar and the interface regions are unaltered, it may be suggested that — under various physiological transitions — their dynamic properties remain more or less constant. This argument reiterates the significance of a proper microviscosity for optimal functioning. An example of stress fluidization and fluidity restoration is shown in Figure 4.

Thermal stress, induced in microorganisms, plants, poikilotherms, and hibernating animals, induces an efficient homeoviscous acclimation which is confined mainly to changes in the degree of unsaturation and length of the phospholipid acyl chains.[69,174,176,177,183] The cholesterol level changes to a lesser, yet significant, extent.[173,179,180] Within a relatively short time after the change in the ambient temperature, the homeostasis is completed and the membrane lipid microviscosity is restored.[69,176-180,183,184] Similar patterns of homeoviscous adaptation occur during hibernation[173,185] and, presumably, in adaptation to pressure.[123]

Treatment of membranes in vitro with drugs, alcohols, or anesthetics increases the lipid fluidity.[30,182,186-192] Upon chronic intake of pharmacologically relevant doses of ethanol, anesthetics, and other drugs, a tolerance is developed. The tolerance is characterized by at least partial restoration of lipid fluidity by increase in cholesterol and saturated fatty acid contents.[179-182,193-196] The adaptation to nutritional stress, an example of which is shown in Figure 4, is similar in its characteristics to temperature acclimation and, presumably, operates via the same mechanism of homeostasis.

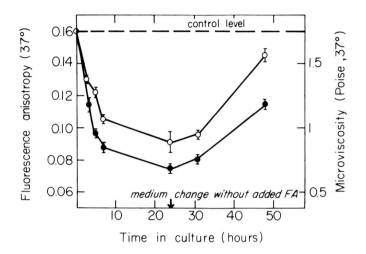

FIGURE 4.   The effect of 40 μg/mℓ oleic acid added to serum-free medium
(○) or medium containing 10% fetal calf serum (●) on the membrane microviscosity of neuroblastoma cells (Clone Neuro-2A®) grown in them. Medium
change after 24 hr without addition of oleic acid triggered a recovery of
membrane microviscosity, as monitored by DPH fluorescence anisotropy.
(Data given in Boostra, J., Nelemans, A. S., Bierman, A., Van Zoelen, E.
J. J., Van der Saag, P. T., and Delaat, S. W., *Biochim. Biophys. Acta*, 662,
321, 1982.)

III. FUNCTIONAL ASPECTS OF LIPID FLUIDITY

Some general features of cell growth and function and its dependence on the plasma
membrane lipid fluidity are now characterized. The rate of cell division is slowed down as
the membrane microviscosity increases. Thus, towards confluency, the cell plasma membrane
acquires a higher microviscosity than during the logarithmic growth phase (Section II.D.2).
Similarly, malignant cells (e.g., leukemic cells) have a lower membrane microviscosity
compared to normal cells. A steady increase in membrane lipid microviscosity takes place
along cellular differentiation (Section II.D.3). Analogously, the membrane microviscosity
of aging cells is at a progressive increase (Section II.D.4). These patterns have the general
feature that growth and vitality of cells are suppressed as the lipids of the cell membrane
become more viscous, and vice versa.

The increase in membrane microviscosity can increase the mechanical strength of the cell
periphery as, for example, in erythrocytes with increased cholesterol level, which become
more resistant to osmotic rupture. This trivial outcome can only partially account for the
above changes in cellular activity upon microviscosity modulation. Undoubtedly, the major
effect of lipid viscosity is on the dynamics of functional units which are embedded in the
lipid matrix. This aspect of the lipid microviscosity is outlined in detail in the following
paragraphs (for reviews, see References 19 and 197).

A. Lateral and Rotational Diffusions of Membrane Proteins

The common biological membrane is highly dense with proteins and can be envisaged as
a loose network of proteins, among which isles of lipid domains are spread. In such a texture,
the diffusion patterns of the proteins are intricate and their correlation with the lipid viscosity
is generally only qualitative. A substantial portion of the membrane proteins appears to be
practically immobile.[19,161,198] Some of the immobile proteins, however, can be mobilized
during metabolic processes, which suggests a physical or chemical association with intra-

cellular cytoskeleton of actin-myosin-type networks.[7,8,198] This type of mobility can be blocked by metabolic inhibitors or at low temperature and is, therefore, markedly different from passive diffusion. The analysis of passive diffusion of membrane proteins suffers not only from the high protein density, but also from the lack of a basic theoretical treatise on two-dimensional diffusion.[13,27] Thus, the correlation between the lipid microviscosity and the rate of lateral diffusion can at this stage be only partially accounted for.[199-201]

Direct experimental measurements (e.g., fluorescence recovery after photobleaching, Reference 202) of lateral passive diffusion of membrane proteins are restricted to distances of over 0.1 μm. At such distances protein-protein encounters act as strong retarding forces, in addition to the cytoskeleton constraint, which is displayed in more than an order of magnitude lower diffusion coefficient of $D < 10^{-10}$ cm^3 · sec^{-1} for integral membrane proteins than what could be estimated from the basic Stoke-Einstein equations. In fact, the only membrane proteins with the expected diffusion constants as yet recorded are rhodopsin in rod outer segments, displaying $D_{25°} \sim 4 \times 10^{-9}$ cm^2 · sec^{-1},[203-205] and receptors on muscle-bled membranes detached from the cytoskeleton ($D \sim 3 \ 10^{-9}$ cm^2 · sec^{-1}).[206] It may, therefore, be expected that upon releasing the cytoskeleton constraints and confining the diffusion to distances of the size of the diffusing proteins (measured, for example, by nonradiative energy transfer), the diffusion coefficient will approach the value recorded for rhodopsin.

Rotational diffusion of membrane proteins, measured by depolarization of phosphorescence[207,208] or photochromic absorption,[199] is, in principle, affected by the immediate environment only. Here, the lipid microviscosity becomes the major retarding force, similar to short-distance lateral diffusion, and the correlation between the rate of rotational diffusion and the lipid microviscosity can be estimated better than in long-range lateral diffusion, yet with possible exceptions.[201,208] The highly anisotropic structure of membranes confines rotational diffusion of proteins to the axis perpendicular to the membrane plane. The transverse modes of rotation around axes parallel to the membrane plane are limited to displacements of small angles and are presumably of negligible contribution.[199,208]

In all cellular activities which are triggered by association of previously separated units, lateral diffusion can become a decisive dynamic parameter. In hormone signals, which are induced by activation of adenylate cyclase through dynamic encounters with the occupied hormone receptor, the overt activity is directly related to the membrane lipid microviscosity.[209] Similar dependence on microviscosity, although not yet demonstrated experimentally, is expected in microaggregation of receptors, which is now believed to be the *unit signal* for signal transduction of a series of hormones[210] and other physiological ligands. An analogous process possibly dictates the associative recognition between membrane antigens and the major histocompatibility complex,[9,211,212] which is believed to act as an immunogenic unit. Similarly, it has been proposed that the trigger mechanism of vision involves microaggregation of the bleached rhodopsin with neighboring rhodopsins to form a transient ion channel.[213-216] In these processes of lateral association, rotational diffusion can provide the proper orientation for matching between the associated proteins.

A proposed model for the involvement of lipid fluidity in the formation of an active aggregate after ligand binding is presented in Figure 5 (see also Section III.D.3).

B. Vertical and Lateral Displacement of Membrane Proteins
1. Lipid-Induced Displacement

The equilibrium position of membrane proteins which are freely diffusible can be regarded as a state of minimum free energy of all the intermolecular interactions of the protein side chains with the ambient lipid and aqueous domains. In this respect, the lipid layer has features of a weak solvent of membrane proteins, due to the considerable difference in dielectric properties between the two. The weak solvent-solute-like interactions between the proteins and lipids implies that it is mostly entropy-driven and is, therefore, highly dependent

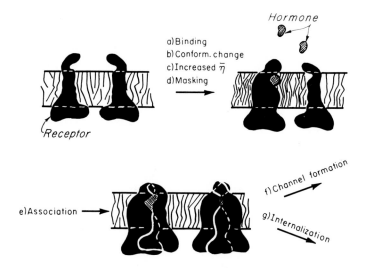

FIGURE 5.   Sequence of events, proposed to follow hormone binding to membrane receptors, which leads to the formation of a unit signal of aggregated receptors.

on the average free volume of the lipid chains, which can be expressed by either density or microviscosity (Section II.C.2).

The inverse proportionality between the lipid-free volume and the microviscosity indicates a decrease in solubilization capacity[23,217-219] of the membrane lipid layer when the microviscosity is increased. In parallel, the net energy of interaction between the proteins and the lipid chains decreases.[220] Thus, a change in equilibrium position of membrane proteins, upon alteration in the lipid viscosity, is directly implied.

For low molecular weight solutes, increase in lipid viscosity will decrease the partitioning into the lipid from aqueous solution.[9,46,217-221] However, when the solute is a protein, with a battery of side chain interactions and with dimensions similar to or greater than the lipid bilayer itself, changes in the lipid microviscosity would mainly result in changes in the degree of exposure of the protein "solute". Alteration in lipid fluidity will change the balance of the opposing interactions with the water and the lipids, and will thus displace the protein to a new equilibrium position. When lipid fluidity is decreased, the new equilibrium position will be of an overall weaker protein-lipid interaction but with correspondingly greater protein-water associations.[220] Thus, the position of the membrane protein may be displaced towards the aqueous phase on either side of the membrane when the lipid microviscosity is increased, and vice versa. In principle, the same concept can also be applied to special amphipathic lipids, which are distinctively different from the bulk membrane lipids. Such lipids may be glycolipids, to which class the blood group antigens belong.

The above concept was promptly verified by spectral,[10,222] chemical,[223] and ultrastructural methods,[224,225] as well as ligand binding,[10,226-228] and was termed as "vertical displacement" of membrane proteins.[222] The compensatory protein displacement ensuing fluidity changes could alternatively operate laterally. Namely, upon increase in microviscosity proteins will associate by "lateral displacement".[224,225,229-231] Reported changes in position and projection of proteins by changes in membrane lipids[229-235] could be accounted for by such vertical or lateral displacement mechanisms. The modulation of any membranal activity, which is associated with protein displacement, is referred to as "passive modulation", since it does not require metabolic energy.[9] Obviously, in immobilized proteins, the displacement by passive modulation will be hindered or even completely abolished.[234,237] Functional aspects

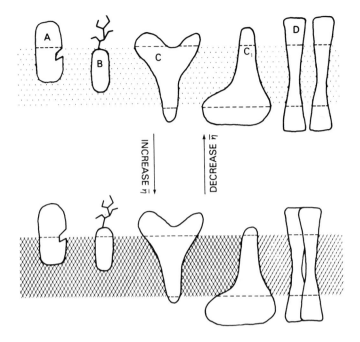

FIGURE 6. Modes of vertical and lateral displacements of membrane proteins mediated by changes in lipid fluidity.

of vertical and lateral displacement of membrane proteins are reviewed in References 19, 238, and 239.

Diffusible membrane proteins and functional lipids can be classified according to their displacement mediated by changes in lipid fluidity, as depicted in Figure 6:

Class A — An integral membrane protein, which is partially exposed on the outer side of the membrane. The figure describes a membrane receptor or enzyme with a binding site near the hydrocarbon-water interface of the lipid bilayer. Small changes in lipid microviscosity can modulate the activity of such sites, especially towards hydrophilic substrates.

Class Ai — An analogous integral membrane protein, which is exposed on the inner side of the membrane.

Class B — A glycoprotein or a glycolipid with a polysaccharide chain, which is exposed on the outer side of the membrane. This site could be an antigen which, by increasing lipid microviscosity, becomes more exposed and thus renders the system more antigenic.

Class Bi — An analogous site, which is exposed on the inner side of the membrane (such sites are probably rare).

Class C — A highly nonsymmetrical cross-membrane protein, with a greater exposure on the outer side of the membrane. The net effect of increase in microviscosity will be an increase in exposure of the protein on the outer surface of the membrane and decrease in its exposure on the inner surface.

Class Ci — A cross-membrane protein, but with a greater exposure on the inner side of the membrane. Increasing microviscosity will increase the exposure of the protein on the inner side of the membrane and will diminish its exposure on the outer side.

Class D — Lateral displacement of a relatively symmetrical cross-membrane protein, where increase in lipid microviscosity will most often increase protein-protein interaction (e.g., receptor-enzyme combination, subunit association, or nonspecific aggregation).

Figures 7 and 8 present examples of electron micrographs, which demonstrate vertical (Figure 7) and lateral (Figure 8) displacements of membrane proteins upon increase in lipid microviscosity.

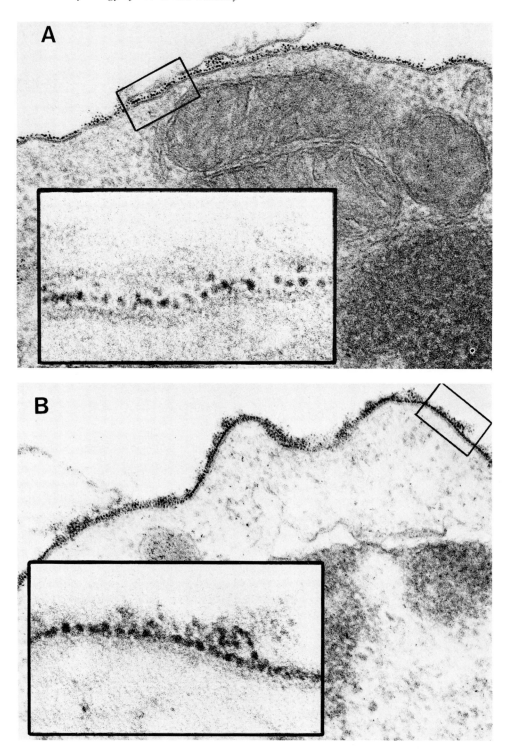

FIGURE 7.   Electron micrographs ($\times 100,000$) of thymocytes from BALB/c mouse before (A) and after (B) rigidification of the membrane lipids by incorporation of cholesteryl hemisuccinate (CHS). Sialic acid residues on the cell surface were first oxidized at $0°$ with sodium periodate and then conjugated with ferritin-hydrazide which is displayed by an electron-densed dot for each separate molecule. As shown, upon incorporation of CHS the sialic acid glycoconjugates are displaced inward and therefore belong to Class Ci. A large portion of these conjugates presumably belong to major histocompatibility antigens. (Communicated by E. Roffman and M. Wilchek.)

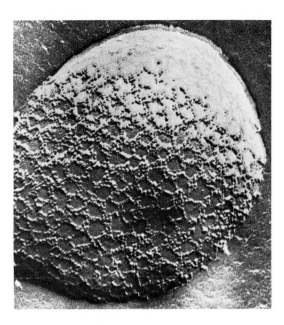

FIGURE 8. Electron micrograph (× 40,000) of freeze fractured cytoplasmic membrane vesicle from *E. coli* grown in the presence of elaidic acid (a lipid rigidifier). The particle distribution (i.e., proteins) indicates local aggregations which are absent in membrane vesicles from *E. coli* grown with no addition of elaidic acid. (Communicated by T. Gulik-Krzywicki.)

The quantitative assessment of the vertical displacement[103] is important in understanding the effect of lipid fluidity on overt physiological functions (Section III.C). For this purpose, one can assume that the site which carries the function is at a quasiequilibrium between operating (C_+) and cryptic (C_-) forms. The fraction of operating sites, α, at any given time and area, is

$$\alpha = C_+/C_0 = 1/[1 + (k_2/k_1)] \tag{17}$$

where C_0 is the total number of operable sites in the system. The kinetic constants, k_1 and k_2, which determine the rate of shuttling between the operative and the cryptic forms, depend on lipid viscosity, but in a complex manner. Qualitatively, while k_1 increases with $\bar{\eta}$, k_2 decreases with $\bar{\eta}$, probably to a similar magnitude, and their ratio should therefore be a power of $\bar{\eta}$:

$$k_1/k_2 \propto \bar{\eta}^m \tag{18}$$

where m is a constant specific to the site. For practical reasons, $\bar{\eta}$ can be expressed in units of $\bar{\eta}_{1/2}$, the specific lipid microviscosity at which only half of the sites are in the C_+ form, which yields:

$$\alpha = 1/[1 + (\bar{\eta}/\bar{\eta}_{1/2})^{-m}] = 1/(1 + \tilde{\eta}^{-m}) \tag{19}$$

The power m is an ''expansion factor'', which characterizes the cooperativity between the lipid microviscosity and the site accessibility. Positive m represents an increase in accessibility with $\bar{\eta}$, while negative m represents decrease in site accessibility. In the general case, where a constant fraction, f, of the sites is in the operating form, without being affected by changes in $\bar{\eta}$, expression will be extended to:

$$\alpha = (1 + f\,\tilde{\eta}^{-m})/(1 + \tilde{\eta}^{-m}) \qquad (20)$$

where $0<f<1$ and $f<\alpha<1$.

The degree of exposure of membrane proteins can also be modulated by perturbations in the electrostatic interactions with the aqueous layer.[240] Alterations in charge distribution and density on the membrane proteins can change the free energy of interactions of protein side chains and their aqueous environment. Such changes can be mediated by subtle pH changes within the physiological range, which may affect the ionization state of residues like histidine, cysteine, and the α-amine. Similar perturbations can be induced by changes in the ambient salt concentration, or in membrane potential. As a result of the electrostatic perturbations, the equilibrium position of the protein can be shifted similar to the process mediated by changes in the lipid fluidity.

2. Ligand-Induced Displacement

Binding of charged ligands can induce marked changes in solubility of proteins in water.[145,241,242] Under extreme conditions of extensive charge neutralization, protein-ligand complexes can even partition from water to organic solvents.[145,241,242] Ligand binding to a membrane protein (i.e., hormone-receptor interaction) could induce such a modulation of charge distribution.[240,243] In addition, the free energy of binding, which — for many specific interactions — is in the order of 10 kcal/mol, can be tunnelled to substantial conformational changes.[244,245]

The ligand-induced changes in charge and conformation of membrane receptors are likely to affect their partitioning characteristics between the aqueous and the lipid domains. Furthermore, the immediate lipid fluidity around the occupied receptor may also change, thus affecting unoccupied receptors (see Section III.D.3). The resultant changes in exposure of both occupied and unoccupied receptors may lead to local aggregations, as depicted in Figure 5. This mechanism may provide the clue to the process of receptor aggregation after ligand-binding, which was observed in several receptor systems, and is now believed to be a general mechanism for triggering of physiological signals[210] (see Section III.A).

3. Conformational Changes

The protein region which is shifted from the hydrophobic core to the hydrophilic domain of the membrane is susceptible to conformational changes similar to that which polypeptide undergoes when transferred from an organic solvent to water.[246] A contribution to the overt conformational changes can also arise from the change in lateral pressure.[121] Thus, upon increase in membrane microviscosity, the number of available sites may increase but their affinity to ligand binding may change as well.[226,246]

C. Lipid Fluidity and the Overt Activity

1. Formulation

The analogy between enzymatic reaction, expressed in terms of Michaelis-Menten kinetics, and membranal activities can serve as a useful tool for quantitative assessment of various cellular functions. In general, the most reliable results are obtained when the external "ligand" (e.g., a substrate for membrane-bound enzyme, a hormone for receptor activation, or a nutrient in carrier-mediated transport) is in excess. Under such conditions, the overt rate of function is maximal (V_{max}) and can be expressed by:[103]

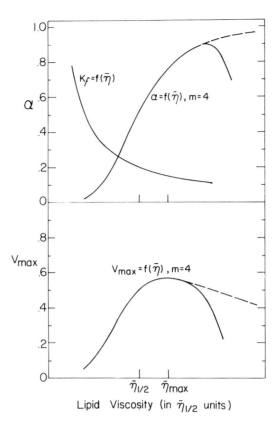

FIGURE 11. Schematic representation of the dependence of
α (Equation 19) and k_f (Equation 22) on the lipid microviscosity
($\bar{\eta}$) when m = 4, and the consequent dependence of V_{max} on
$\bar{\eta}$. The broken lines represent the case when no irreversible loss
of operating units (e.g., shedding) occurs.

apparent degree of accessibility of receptors and its response to fluidity changes.[226,227,236,253,254]
In addition, it provides an assessment of the total receptor capacity which is stored in the
membrane (Section III.C.5). A summary of such evaluations is given in Table 3. In general,
it appears that receptors of neurological function are the most sensitive to such changes,
which presumably reflects part of their operational mechanism. An example which presents
a simultaneous determination of five different neurological receptors is shown in Figure 12.
It clearly demonstrates the diversity in response to changes in lipid fluidity among the
membrane receptors. The case of the β-adrenergic receptor (see Table 3) indicates that in
different membranes the same receptor can adopt different displacement characteristics,
presumably in accordance with its local functional patterns.

In long-term changes of membrane lipid microviscosity, as during aging (Section II.D.4),
the total number of receptors may change by various metabolic processes as a result of
overexposure. Reported data on modulation of receptors in aging are summarized in Table
4.

2. Receptor Modulation in Aging

The level of operating membrane receptors and tissue responsiveness declines simultaneously with aging.[264,265] The progressive increase in lipid microviscosity with aging (Section
II.D.4) predicts that at least part of the overt reduction in receptor function is due to a
reduced capacity and plasticity (see Section III.C.5). The changes in receptor functions with

Table 3
ACCESSIBILITY MODULATION OF MEMBRANE RECEPTORS BY LIPID FLUIDITY

Ligand	Target membranes	The effect of increased microviscosity on receptor accessibility	Estimated degree of physiological accessibility (%)	Ref.
Serotonin	Rat brain	Marked increase	10—20	226, 236, 253
Opiate	Rat brain	Increase	50—70	253, 254
β-Adrenergic	Rat liver	Increase		70
	Rat reticulocytes	Decrease		219
	Turkey erythrocytes	Unchanged		209
Thrombin	Human platelets	Increase	50	255
5-Hydroxy indole acetaldehyde	Dog brain	Decrease		256
TSH	Human thyroid	Decrease		257
ACTH	Human thyroid	Decrease		257
Fc	Rat spleen cells	Unchanged		258
	Leukemia cells	Increase		259
Transferrin	Human erythrocytes	Decrease	50	227
	Friend cells	Decrease	30	163
Concanavalin-A	L-cells	Decrease		260
Immunoglobulin	Rabbit splenocytes	Increase		261
Insulin	Fetal rabbit microsomes	Increase		262
Angiotensin II	Bovine adrenal glomerulosa	Decrease		263

Table 4
CHANGES IN ACCESSIBILITY AND CAPACITY OF MEMBRANE RECEPTORS UPON AGING

Receptor	Membrane	Accessibility	Capacity	Ref.
Serotonin	Rat brain	Increase	Decrease	226, 236, 253, 266, 267
Opiate	Rat brain	Unchanged	Slight decrease	253, 254
β-Adrenergic	Human brain	Decrease		268, 269
	Rat adipocytes	Decrease		353
	Rat erythrocytes	Decrease		270
Acetylcholine	Rat brain	Decrease		271
Insulin	Human fibroblasts	Increase		272

aging, which are summarized in Table 4, could, to a substantial extent, be mediated by the decrease in lipid fluidity.

3. Fluidity Changes by Ligand Binding

Specific association between a membrane receptor and its ligand is generally of a high affinity and is, therefore, characterized by a relatively high free energy of interaction. This energy may mediate significant conformational changes in the occupied receptor,[273] which can change the packing density of the neighboring lipid layer. The resulting fluidity changes are expected to fade away from the occupied receptor, but opposing fluidity gradients from other occupied receptors will direct the system to a new steady-state level of lipid fluidity. Recently, independent reports, which are summarized in Table 5, have described such lipid fluidity changes.

Table 7
MODULATION OF CARRIER-MEDIATED TRANSPORT ACTIVITIES BY CHANGES IN THE MEMBRANE LIPID FLUIDITY

Transported ligand	Membrane	Lipid manipulation	Corresponding change in microviscosity	Change in activity	Ref.
Na+, K+	Human erythrocytes	Increased cholesterol	Increase	Decrease	57, 296
	Guinea pig erythrocytes	Increased cholesterol	Increase	Decrease	294
	Human erythrocytes	Decreased cholesterol	Decrease	Increase	57, 296
	Human erythrocytes	Increased lecithin	Decrease	Increase	308
Glucose	Human erythrocytes	Decreased cholesterol	Decrease	Decrease	103, 295
	3T3 Fibroblasts	Increased cholesterol	Increase	Increase and then decrease	103
	3T3 Fibroblasts	Decreased cholesterol	Decrease	Decrease	103
Galactose	Human erythrocytes	Anesthetic alcohols	Decrease	Decrease	307
L-Arabinose	Human erythrocytes	Increased cholesterol	Increase	Increase	292
L-Lactate	Human erythrocytes	Decreased cholesterol	Decrease	Decrease	292
	Human erythrocytes	Increased cholesterol	Increase	Increase	292
Uridine	Human erythrocytes	Decreased cholesterol	Decrease	Decrease	292
	Human erythrocytes	Anesthetic alcohols	Decrease	Decrease	307
Thiourea	Guinea pig erythrocytes	Increased cholesterol	Increase	Decrease	294

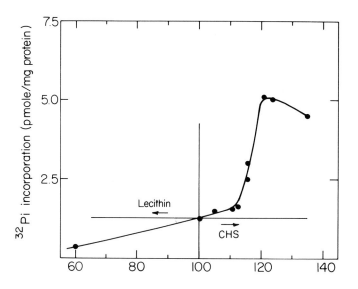

FIGURE 13. The endogenous phosphorylation of a membrane protein ("Protein C") in P₂M membranes from mouse forebrain (see Figure 12) after lipid rigidification with CHS and fluidization with lecithin. The apparent phosphorylation is the net of the rate of phosphokinase activity and the opposing rate of dephosphorylation by phosphatases. (From Hershkowitz, M., Heron, D., Samuel, D., and Shinitzky, M., *Prog. Brain Res.*, 56, 419, 1982. With permission.)

the membrane, and are, by and large, freely diffusible. As such, their degree of exposure could be manipulated by changes in the membrane lipid microviscosity both in vivo and in vitro.[10,228,309-311] Again, as in the case of membrane receptors or enzymes, the increase in antigen exposure, which may be correlated with increase in antigenicity, may not necessarily correspond to an increase in immunogenicity. The latter refers to the capacity for eliciting an immune response, and is a combination of antigenicity and the rate of antigenic processing to form the unit of immunogenic signal. Similar to the other membranal functions, a maximal immunogenicity may be associated with a specific lipid microviscosity, above which the immunogenicity may decline, in spite of an apparent increase in antigen exposure (see Section III.C.4).

1. Established Antigens

The degree of exposure of blood-group antigens can be readily modulated by changing the membrane lipid fluidity in human erythrocytes.[10,228] This can be achieved by depletion of cholesterol, for fluidization, or by incorporation of cholesterol or its hydrophilic ester cholesteryl hemisuccinate (CHS), for rigidification.[228] A demonstration of the increase of Rh(D) antigens upon cholesterol enrichment is shown in Figure 10. The figure also demonstrates that, upon additional lipid rigidification by hydrostatic pressure, most of the Rh(D) antigens are shed off (Section III.C.3) and can then be collected from the supernatant.[228] In an analogous example, Figure 14 presents the modulation of Θ-antigens on mouse T-lymphoma cells. It is interesting that antigens of the histocompatibility complex behave in a converse manner to that of the Rh or Θ antigens.[311] Increase in lipid microviscosity seems to mask these antigens reversibly. The masking of sialic acids upon incorporation of cholesterol, which is shown in Figure 7, could be correlated with this observation, since a large portion of the membrane sialic acids is believed to be associated with the major histocompatibility complex. When further verified, this notion could suggest a lipid treatment for suppression of these antigens (e.g., prior to tissue transplantation).

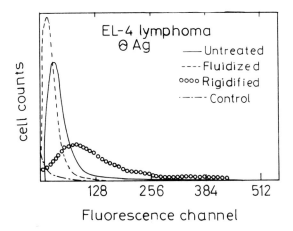

FIGURE 14. Fluorescence activated cell sorting profiles of Θ-antigen accessibility in untreated (——), cholesterol-enriched (OOOO) and cholesterol-depleted (---) EL-4 mouse lymphoma cells. After lipid modification the cells were reacted with anti-Θ antibodies and then with fluoresceinated rabbit anti-mouse IgG. In the control cells (-··-) the cells were directly reacted with the fluorescent anti-IgG without reacting with the anti-Θ. The peak of the profiles represents approximately the average levels of accessible Θ-antigens, whereas its width indicates the apparent heterogeneity in Θ-antigen accessibility in the tested cell population.

Among the in vivo processes in which the microviscosity of cell membranes is increased, aging is of relevance to possible changes in antigenic expression (see Section II.D.4). In lymphoid cells of old animals[170] and old men,[169] the apparent increase in lipid microviscosity correlates well with increase in cholesterol level. It is, therefore, expected that antigens in the aged cell could become more exposed but less mobile. Also, the additional surface area of cells,[7,41,258] which is presumably due to the excess cholesterol[41] could reduce the apparent number of antigens per surface area. The average number per cell of the antigen AgB in rat splenocytes was found to be greater in cells from old than from young animals, but the density of antigens was estimated to be higher in cells from young animals.[258]

2. Tumor-Associated Antigens

It is now well accepted that malignant cells bear specific antigens which, in principle, can elicit humoral or cellular immune response.[312] The shortage in availability of specific and well-defined antibodies against these putative antigens entails the use of indirect functional tests for monitoring changes in antigenicity of tumor cells. The most relevant is prophylactic immunization with irradiated syngeneic tumor cells of modulated antigenicity. The degree of elicited immunity can be assessed by survival after challenge with untreated viable tumor cells. Experiments of this kind have indeed demonstrated that, upon increase of the cell membrane microviscosity by CHS, for example, the immunity induced by the "vaccine" made of these cells is markedly stronger than that induced by the untreated cells.[313-315] Analogously, delayed cutaneous hypersensitivity tests (skin reaction) were performed in cancer patients with their autologous irradiated tumor cells.[314,316,317] In most cases, only after membrane rigidification (with CHS), a strong skin reaction was elicited, again indicating increase in antigenicity.[316,317]

3. Clinical Implications

The ''fortification'' of ''weak'' antigens, by manipulation of the membrane microviscosity, has the important advantage of being nontoxic, nonadversive, and maintaining the overall texture of surface determinants. Diagnostically, this method could facilitate the detection of buried antigens like the cold agglutinin (I-antigen) in human erythrocytes.[40] Clinically, it opens up a new approach for augmentation of antigenicity as for cancer active immunotherapy, or suppression of antigenicity as for tissue transplantation (see Section IV). These possibilities have not yet entered into routine medical practice.

Another intriguing aspect of passive antigenic modulation relates to autoimmune diseases which originate from overexposure of normal membrane constituents. As a result of an abnormal membrane microviscosity, antigenic residues which are naturally masked by the membrane become exposed and faultily recognized as nonself. This possibility is pertinent to aging when cell membranes are progressively rigidified (see Section II.D.4).

G. Normal and Disordered Complex Functions

Most physiological activities that are confined to the cell membrane are a complex combination of protein accessibility, turnover rates, and lateral association and dissociation of the involved membrane components. Since each of these functions may depend on the lipid microviscosity, the overt cellular function can be modulated by changes in the membrane lipid microviscosity in a complex manner, which at the current state can be analyzed only qualitatively. An example of the dependence on microviscosity of a complex membranal function is shown in Figure 13. To date, the most studied complex physiological functions are immunological reactions. Slight to moderate cholesterol enrichment of lymphocytes was found to potentiate lymphocyte responses;[318-320] higher enrichment with cholesterol was found to markedly inhibit lymphocyte activity.[320,321] The latter condition may be pertinent to the suppression of immune competence with age[322] and under nutritional stress.[323,324] Since cholesterol depletion of lymphocytes was also found to inhibit mitogenic stimulation,[320] it is apparent that the overt lymphocyte stimulation has an optimum microviscosity for maximal response[325,326] (Section III.C.4). It should be noted, however, that modulation of cell membrane fluidity with fatty acids did not result in any substantial effect on mitogenic response.[327,328] Yet, the effect of cholesterol on the activities of macrophages[329] and monocytes[330] is similar to that on lymphocytes[330,331] and, in general, it appears that cholesterol could act as a natural modulator of immune responses[331,332] and tumor development.[165,333] Other complex functions, which were found to correlate with membrane lipid fluidity, were electrical activity,[139] cell agglutination,[162] cell fusion,[334] viral infectivity,[55,335] sperm fertility,[336-339] and egg fertilization.[340]

The two main factors which can change the inherent lipid fluidity of cell membranes are the lipid composition of the serum, which to a large extent is determined by the liver function and to a lesser extent by the diet, and the intracellular lipid metabolism. The various processes of adaptation to stress (Section II.D.5) indicate that in a normal tissue the intracellular lipid metabolism, which is associated with homeostatic pathways, is sufficient to adjust any lipid alteration induced by the serum lipids. In liver malfunction[57] and overnutrition,[323] the abnormal lipid composition of the serum may reach a level which can override the intracellular homeostatic mechanisms and lead to a chronic change in cell membrane lipid composition and fluidity.[40,41,57] Therefore, a secondary outcome of various diseases and disorders, in which the composition of serum lipids is changed, could be a change in the lipid fluidity of cell membranes.[9,40-42,57,80-82,84,85,167,341-344] It is interesting that, in all of these cases, the cell membranes become more viscous mostly due to an increase in cholesterol/phospholipids[9,40-42,57,84,85,167,341-344] or decrease in lecithin/sphingomyelin.[80-82] Similar changes occur in aging and are predominantly caused by the weakened intracellular homeostatic lipid metabolism (Section II.D.4). Diseases with more fluid cell membranes are leukemia[345,346] and Chediak-Higashi Syndrome.[347,348]

The secondary changes in cell membrane lipid fluidity, which were described above, could be manifested in various abnormalities of complex physiological functions. It could also be the etiology of psychological and behavioral deviations. These considerations also apply to the primary phase of drug intake, where cell membranes are instantaneously fluidized, and to the phase of drug withdrawal, when the fluidity-adapted membranes become hyperviscous. [181,182,189,194]

IV. TOWARDS MEMBRANE ENGINEERING

Except for a small portion of functional lipids (e.g., gangliosides), the majority of the membrane lipids serves as an assembly assigned to provide a defined fluid environment for optimal function, which by and large is carried out by the membrane proteins. The correspondence between membranal function and lipid fluidity seems to operate along the sub-macroscopic scale of microviscosity (Section II.A) and, to a much lesser extent, according to a specific lipid composition. Thus, as outlined in Section II.D, stress conditions which affect lipid fluidity can trigger a series of biosynthetic pathways which are directed towards restoration of the lipid fluidity (Section II.D.5). However, in some cases where the inherent lipid metabolism and membrane proteins are still in the normal range, the rate of fluidity adjustment may be too slow to compete with the rate of fluidity distortion and the overt membrane fluidity may appear at a level which deviates from normal. This instance may be pertinent to the early phase of aging (Section II.D.4), obesity,[323,349] hyperlipidemia,[41] or hypertension.[342-344] Yet, as long as the composition of the functional sites (i.e., the proteins) remains normal, outside intervention with specific lipid mixtures designed to readjust the lipid fluidity (but not necessarily the lipid composition) may restore cellular function.

In all the cases mentioned above, the apparent membrane lipid microviscosity is above normal ("hyperviscous"). Another interesting case of hyperviscous cell membranes is found upon drug withdrawal in drug addicts. The presence of addictives, like alcohol, morphine, or anesthetics, is known to fluidize membrane lipids by direct incorporation or indirectly by binding to receptors.[182,187-189,192,195] Chronic intake of such "fluidizers" promotes adaptive adjustment of membrane lipid fluidity, generally by increasing cholesterol content.[181,182,189,194,195] Upon abrupt withdrawal of the drug after prolonged chronic intake, cell membranes, in the brain in particular, instantaneously become hyperviscous with all the physiological consequences relating to changes in protein exposure and mobility (Section III). It is very likely that the well-known drug withdrawal syndrome of addicts originates from this hyperviscous state of the brain membranes.[189] A schematic description of the modulations in membrane fluidity and drug receptors at the various phases of drug addiction is shown in Figure 15.

The obvious candidate for membrane fluidization both in vitro and in vivo is lecithin from natural sources (e.g., egg yolk), which can fluidize membranes either by extracting excess cholesterol or by incorporation into the membrane, either passively or by exchange (see Section II.B). The efficacy of the fluidization activity of lecithin depends primarily on the physical state at which it is presented to the membranes. This can be in the form of liposomes,[42,57] enriched serum,[349] or a complex with polyvinyl-pyrrolidone (PVP) which acts as a carrier.[55,103,313] For in vivo use, each of these forms has some disadvantages. Recently, we developed in our laboratory a potent lipid mixture, designated as Active Lipid (AL), for membrane fluidization both in vitro and in vivo. This mixture is composed of about 70% neutral glycerides, 20% lecithin, and 10% phosphatidyl ethanolamine, all from hen egg yolk which, when dispersed in water, forms micelle-like structures where the neutral lipids form the core and the phospholipids are distributed on its surface as a monolayer.[351] This structure seems to render the participating phospholipids with high fluidization capacity. Preliminary reports on restoration of brain functions in old mice and alleviation of abstinence

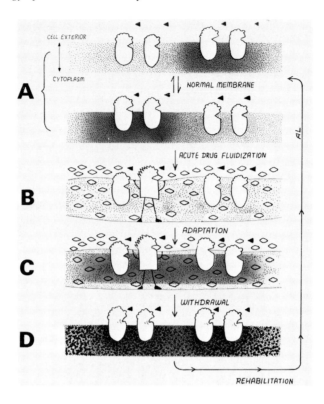

FIGURE 15. Membrane fluidity phases in drug addiction. A — normal membrane where the drug receptor is about 50% accessible for binding; B — acute lipid fluidization, upon the initial drug intake; C — adaptively adjusted membrane fluidity under the chronic presence of the drug; D — the hyperviscous state after abrupt drug withdrawal which presumably mediates the abstinence syndromes. (From Heron, D. S., Shinitzky, M., and Samuel, D., *Eur. J. Pharmacol.*, 83, 253, 1982. With permission.)

syndromes in morphine-addicted mice[189] indicate a great potential of AL for various in vivo treatments.

For membrane rigidification, cholesterol,[33,41,42,55,321] or more efficiently, one of its hydrophilic esters[103,263,313,352] can be used. The method for representation can be in liposomes,[33,42,57] enriched serum,[349] or PVP dispersion.[103,313] Rigidification of membrane lipids may be of great potential for exposure of tumor-associated antigens,[313,316] which may render tumor cells with specific immunogenicity. This approach can elicit antitumor immune reactivity both in vivo and in vitro Similarly, it may be applied for suppression of the major histocompatibility complex before allogeneic tissue transplantation (Section III.F.3).

The lung surfactant is a special membrane where lipid manipulation can be carried out externally through inhalation. The lipid composition of the lung surfactant is of a defined fluidity and mechanical properties which, on the one hand, control the diffusion of O_2 and CO_2 and, on the other hand, support the expansion and relaxation of the alveoli in respiration.[119] Improper lipid composition of the lung surfactant is displayed by microviscosity above normal and is a characteristic of premature newborns[79,83] or of allergic alveolitis in adults.[52] Artificial lung surfactants, which can be applied externally, are currently being tested for restoration of normal respiration in these disorders.

Membrane lipid engineering in disorders associated with an abnormal membrane fluidity has only recently attracted attention. This can be performed either externally, or through special diets, and in more severe cases by infusion. Since no obvious adverse effects are expected in such treatments, it is likely that this approach may become a key method in clinical treatments in the near future.

REFERENCES

1. **Adam, G. and Delbrück, M.,** Reduction of dimensionality in biological diffusion processes, in *Structural Chemistry and Molecular Biology,* Rich, A. and Davidson, W., Eds., W. H. Freeman, San Francisco, 1968, 198.
2. **Goldman, R. and Katchalski, E.,** Kinetic behavior of a two-enzyme membrane carrying out a consecutive set of reactions, *J. Theor. Biol., 32,* 243, 1971.
3. **Richter, P. H. and Eigen, M.,** Diffusion-controlled reaction rates in spheroidal geometry: application to repressor-operator association and membrane-bound enzymes, *Biophys. Chem.,* 2, 255, 1974.
4. **Berg, H. C. and Purcell, E. M.,** Physics of chemoreception, *Biophys. J.,* 20, 193, 1977.
5. **Minton, A. P.,** The bivalent ligand hypothesis: a quantitative model for hormone action, *Mol. Pharmacol.,* 19, 1, 1981.
6. **Hackenbrock, C. R.,** Lateral diffusion and electron transfer in the mitochondrial inner membrane, *Trends Biochem. Sci.,* 6, 151, 1981.
7. **Edelman, G. M.,** Surface modulation in cell recognition and cell growth, *Science,* 192, 218, 1976.
8. **Nicolson, G. L.,** Transmembrane control of the receptors on normal and tumor cells. I. Cytoplasmic influence over cell surface components, *Biochim. Biophys. Acta,* 457, 57, 1976.
9. **Shinitzky, M.,** The concept of passive modulation of membrane responses, *Dev. Cell Biol.,* 4, 173, 1979.
10. **Shinitzky, M. and Souroujon, M.,** Passive modulation of blood-group antigens, *Proc. Natl. Acad. Sci. U.S.A.,* 76, 4438, 1979.
11. **Luzzati, V. and Tardieu, A.,** Lipid phases: structure and structural transitions, *Annu. Rev. Phys. Chem.,* 25, 79, 1974.
12. **Gavish, B.,** Position-dependent viscosity effects on rate coefficients, *Phys. Rev. Letts.,* 44, 1160, 1980.
13. **Saffman, P. G. and Delbruck, M.,** Brownian motion in biological membranes, *Proc. Natl. Acad. Sci. U.S.A.,* 72, 3111, 1975.
14. **Dilger, J. P., McLaughlin, S. G. A., McIntosh, T. J., and Simon, S. A.,** The dielectric constant of phospholipid bilayers and the permeability of membranes to ions, *Science,* 206, 1196, 1979.
15. **Fragata, M. and Bellemare, F.,** Model of singlet oxygen scavenging by α-tocopherol in biomembranes, *Chem. Phys. Lipids,* 27, 93, 1980.
16. **Rothman, J. E. and Lenard, J.,** Membrane asymmetry, *Science,* 195, 743, 1977.
17. **Cogan, U. and Schachter, D.,** Asymmetry of lipid dynamics in human erythrocytic membranes studied with impermeant fluorophores, *Biochemistry,* 20, 6396, 1981.
18. **Shinitzky, M. and Barenholz, Y.,** Fluidity parameters of lipid regions determined by fluorescence polarization, *Biochim. Biophys. Acta,* 525, 367, 1978.
19. **Shinitzky, M. and Henkart, P.,** Fluidity of cell membranes: current concepts and trends, *Int. Rev. Cytol.,* 60, 121, 1979.
20. **Seelig, J. and Seelig, A.,** Lipid conformation in model membranes and biological membranes, *Q. Rev. Biophys.,* 13, 19, 1980.
21. **Smith, I. C. P.,** Organization and dynamics of membrane lipids as determined by magnetic resonance spectroscopy, *Can. J. Biochem.,* 57, 1, 1979.
22. **McConnell, H. M.,** Molecular motion in biological membrane, in *Spin Labeling: Theory and Applications,* Berliner, L. J., Ed., Academic Press, New York, 1976, 525.
23. **Shinitzky, M. and Yuli, I.,** Lipid fluidity at the submacroscopic level: determination by fluorescence polarization, *Chem. Phys. Lipids,* 30, 261, 1982.
24. **Shinitzky, M. and Barenholz, Y.,** Dynamics of the hydrocarbon layer in liposomes of lecithin and sphingomyelin containing dicetylphosphate, *J. Biol. Chem.,* 249, 2652, 1974.
25. **Cogan, U., Shinitzky, M., Weber, G., and Mishida, T.,** Microviscosity and order in the hydrocarbon region of phospholipid and phospholipid-cholesterol dispersions determined with fluorescence probes, *Biochemistry,* 12, 521, 1973.
26. **Shinitzky, M., Dianoux, A. C., Gitler, C., and Weber, G.,** Microviscosity and order in the hydrocarbon region of micelles and membranes determined with fluorescence probes. I. Synthetic micelles, *Biochemistry,* 10, 4335, 1971.

27. **Galla, H. J. and Sackmann, E.,** Lateral diffusion in the hydrophobic region of membranes: use of pyrene excimers as optical probes, *Biochim. Biophys. Acta,* 339, 103, 1974.

28. **Devaux, P. and McConnell, H. M.,** Lateral diffusion in spin labelled phosphatidylcholine multilayers, *J. Am. Chem. Soc.,* 94, 4475, 1972.

29. **Heyn, M. P., Cherry, R. J., and Dencher, N. A.,** Lipid-protein interactions in bacteriorhodopsin: dimyristoyl-phosphatidylcholine vesicles, *Biochemistry,* 20, 840, 1981.

30. **Jahring, F.,** Structural order of lipids and proteins in membranes: evaluation of fluorescence anisotropy data, *Proc. Natl. Acad. Sci. U.S.A.,* 76, 6361, 1979.

31. **Van Blitterswijk, W. J., Van Hoeven, R. P., and Van der Meer, B. W.,** Lipid structure order parameters in biomembrane derived from steady-state fluorescence polarization measurement, *Biochim. Biophys. Acta,* 644, 323, 1981.

32. **Chapman, D.,** Recent studies of lipids, lipid cholesterol and membrane system, in *Biological Membranes,* Vol. 2, Chapman, D. and Wallach, D. F., Eds., Academic Press, New York, 1973, 91.

33. **Shinitzky, M. and Inbar, M.,** Microviscosity parameters and protein mobility in biological membranes, *Biochim. Biophys. Acta,* 433, 133, 1976.

34. **Bach, D. and Chapman, D.,** Calorimetric studies of biomembranes and their molecular components, in *Biological Microcalorimetry,* Beezer, A. E., Ed., Academic Press, London, 1980, 275.

35. **Huang, C.,** Configurations of fatty acyl chains in egg phosphatidylcholine-cholesterol mixed bilayers, *Chem. Phys. Lipids,* 19, 150, 1976.

36. **Huang, C.,** A structural model for the cholesterol-phosphatidylcholine complexes in bilayer membranes, *Lipids,* 12, 348, 1977.

37. **Suckling, K. E., Blair, H. A. F., Boyd, G. S., Graig, I. F., and Malcolm, B. R.,** The importance of the phospholipid bilayer and the length of the cholesterol molecule in membrane structure, *Biochim. Biophys. Acta,* 551, 10, 1979.

38. **Dekruijff, B., Van Dijck, P. W. M., Demel, R. A., Schuijff, A., Brant, F., and Van Deenen, L. L. M.,** Non-random distribution of cholesterol in phosphatidylcholine bilayers, *Biochim. Biophys. Acta,* 356, 1, 1974.

39. **Demel, R. A. and Dekruijff, B.,** The function of sterols in membranes, *Biochim. Biophys. Acta,* 457, 109, 1976.

40. **Cooper, R. A.,** Abnormalities of cell membrane fluidity in the pathogenesis of disease, *N. Engl. J. Med.,* 197, 371, 1977.

41. **Cooper, R. A.,** Influence of increased membrane cholesterol on membrane fluidity and cell function in human red blood cells, *J. Supramol. Struct.,* 8, 413, 1978.

42. **Cooper, R. A., Leslie, M. H., Fischkoff, S., Shinitzky, M., and Shattil, S.,** Factors influencing the lipid composition of red cell membranes *in vitro:* production of red cells possessing more than two cholesterols per phospholipid, *Biochemistry,* 17, 327, 1978.

43. **Pal, R., Barenholz, Y., and Wagner, R. R.,** Effect of cholesterol concentration on organization of viral and vesicle membranes, probes by accessibility to cholesterol oxidase, *J. Biol. Chem.,* 225, 5802, 1980.

44. **Beiderman, B., Whitney, J. O., and Thaler, M. M.,** Regulation of cell surface microviscosity during the cell cycle in hepatoma, *Gastroenterology,* 76, 1275, 1979.

45. **Bloj, B. and Zilversmit, D. B.,** Asymmetry and transposition rates of phosphatidyl choline in rat erythrocyte ghosts, *Biochemistry,* 15, 1277, 1976.

46. **Chen, L., Lund, D. B., and Richardson, T.,** Essential fatty acids and glucose permeability of lecithin membranes, *Biochim. Biophys. Acta,* 225, 89, 1971.

47. **Cheng, S. and Levy, D.,** The effect of cell proliferation on the lipid composition and fluidity of hepatocyte plasma membranes, *Arch. Biochem. Biophys.,* 1976, 424, 1979.

48. **Jackson, R. L., Morrisett, J. D., and Gotto, A. M., Jr.,** Lipoprotein structure and metabolism, *Physiol. Rev.,* 56, 259, 1976.

49. **Mahley, R. W. and Inneranity, T. L.,** Lipoprotein receptors and cholesterol homeostasis, *Biochim. Biophys. Acta,* 737, 197, 1983.

50. **Graham, J. M. and Green, C.,** The properties of mitochondria enriched *in vitro* with cholesterol, *Eur. J. Biochem.,* 12, 58, 1970.

51. **Jonas, J.,** Nuclear magnetic resonance at high pressure, *Annu. Rev. Phys. Chem.,* 26, 167, 1975.

52. **Jouanel, P., Motta, C., Brun, J., Molina, C., and Dastugue, B.,** Phospholipids and microviscosity study in broncho-alveolar lavage fluid from control subjects and from patients with extrinsic allergic alveolitis, *Clin. Chem. Acta,* 115, 211, 1981.

53. **Kawasaki, Y., Wakayama, N., Koike, T., Kawai, M., and Amano, T.,** A change in membrane microviscosity of mouse neuroblastoma cells in association with morphological differentiation, *Biochim. Biophys. Acta,* 509, 440, 1978.

54. **Lange, Y., Cohen, C. M., and Posnansky, M. J.,** Transmembrane movement of cholesterol in human erythrocytes, *Proc. Natl. Acad. Sci. U.S.A.,* 74, 1538, 1977.

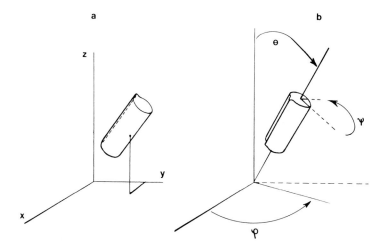

FIGURE 1. Positional (A) and angular (B) coordinates of a rigid body.

$$D_T = kTb_T \; ; \qquad D_R = kTb_R \qquad (2)$$

where k is Boltzmann's constant, T is the absolute temperature, and b is the mobility defined as the velocity (or angular velocity) produced by steady unit force (or torque).[8,9] Note that D_T has the dimension of distance squared times frequency and D_R has the dimension of frequency. A particle having cylindrical symmetry has two (different) rotational constants, $D_\parallel$ for rotations about the symmetry axis and $D_\perp$ for rotations about an axis perpendicular to the symmetry axis; $D_\perp$ is sometimes called the wobbling diffusion constant. In ordinary (isotropic, three-dimensional) fluids the diffusion constants are proportional to the fluidity. This simple relation generally does not hold in membranes (see Section III.A below). However, the diffusion constants for membrane proteins and lipids do have a clear meaning and can be measured by spectroscopic techniques.[5,6] In such a measurement a correlation function is obtained which is essentially a product of a property at time zero and the same property at time t, averaged over all molecules or over all possible starting times. A typical example for such a correlation function is

$$g(t) = <P_2(\cos\theta_0)P_2(\cos\theta_t)> \qquad (3)$$

where the brackets denote an average and P_2 is the second Legendre polynomial, defined as:

$$P_2(x) = \frac{3}{2} x^2 - \frac{1}{2} \qquad (4)$$

The θ_0 and θ_t are the polar angles of the molecular axes at time 0 and t, respectively.

The behavior of g(t) is sketched in Figure 2. This could be constructed by calculating $P_2(\cos\theta(t_0))P_2(\cos\theta(t_0+t))$ for many different t_0 and t values and then averaging over all t_0. The slope of g(t) contains information on the wobbling diffusion constant since it is proportional to $1/D_\perp$. The correlation function g(t) determines the degree of fluorescence depolarization,[10-12] related to the rotational diffusion of probes in membranes.[13] The limiting value of g(t) contains information on another aspect of the motion: the orientational order.

B. Order

The membrane lipid layer is an ordered fluid, where orientation perpendicular to the

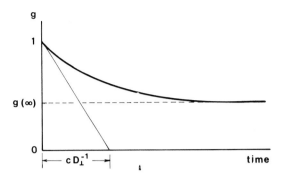

FIGURE 2.   The variation of the correlation function g with time. The starting value equals unity, the limiting value is denoted as g (∞). The slope is proportional to $D_\perp/c$, where $D_\perp$ is the wobbling diffusion constant and c is a constant.

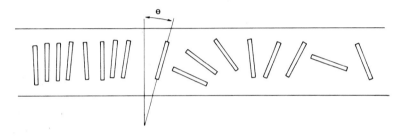

FIGURE 3.   Lipid distribution with high order (left) and low order (right).

membrane plane is more probable than in-plane. This is shown in Figure 3, where the bars denote the axes of the lipids. Two situations are shown, high and low orientational order. The order parameter S, which defines such a distribution, is the average of $P_2(\cos\theta)$:

$$S = <P_2(\cos\theta)> = <(3\cos^2\theta - 1)/2 > \tag{5}$$

S varies between $-1/2$ and 1. These limiting cases have the following meaning: when S $= 1$, all the angles θ equal zero, and the molecules are all exactly perpendicular to the membrane plane. When S $= -1/2$, all the angles θ are equal to 90° and the molecular axes are randomly distributed in the plane. Another interesting limit is S $= 0$, which corresponds to a complete random distribution of molecular axes. In practice it can be assumed that the distribution of orientations in the fluid phase depends only on θ, and not on ϕ or ψ. Note that the long-time behavior of the correlation function g(t) defined above depends on order, because for t $= \infty$, we have

$$g(\infty) = <P_2(\cos\theta_0)><P_2(\cos\theta_\infty)> = S^2 \tag{6}$$

The average of $P_1(\cos\theta) = \cos\theta$ is essentially zero for lipids, because the lipids exchange between the inner and outer monolayer (this motion was termed as flip-flop), and the number of lipids in the two layers is approximately equal. However, the lipid flip-flop time is rather long (in the order of several hours for the erythrocyte membrane at 20°C,[14] and much longer for sphingomyelin in influenza virus membranes[15]). No appreciable flip-flop mobility has yet been observed for membrane-bound proteins, so that natural membranes are in fact asymmetric with respect to lipids and proteins.[15]

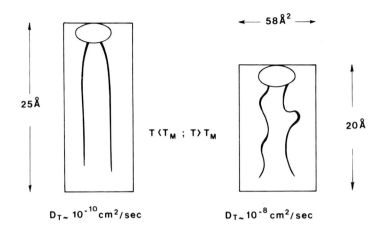

FIGURE 4.   Specific volume and lateral diffusion data for a dipalmitoyllecithin membrane above and below the chain melting transition $T_M = 42°C$. The change in volume is only 3%, but the change in area is 20%.

C. Packing

The viscosity of ordinary fluids correlates with the specific volume. In a series of aliphatic hydrocarbons and similar organic fluids, where the main cohesive forces are van der Waals' attractions, this correlation is given by the empirical Batschinski relation,

$$\eta = B/\Delta V \tag{7}$$

where B is a material constant with the dimension of energy times frequency and ΔV is the free volume of a molecule; that is, the volume minus the volume of the molecular backbone.[16,17] It is plausible that a loose packing will facilitate molecular motion in a membrane as well. In fact, the combination of specific volume data with lateral diffusion data for dipalmitoylphosphatidylcholine model membranes suggests that diffusion correlates with specific area.[17,18] These data are depicted in Figure 4. In this system, phase transition is associated with increase in specific area and decrease in thickness. Therefore, volume changes are small compared to area changes in this model membrane. The lateral diffusion constants displayed in Figure 4 refer to a lipid probe and have been determined by a fluorescence technique.[19] In biologically more relevant systems, volume changes are expected to be close to the changes in specific area.[17]

D. Permeability

Permeability across a lipid bilayer is a combination of partitioning and passive diffusion, both of which depend on lipid fluidity. The permeability of nonelectrolytes could thus be used as a means to study membrane fluidity.[20]

III. RELATIONS BETWEEN FLUIDITY PARAMETERS

In this interesting area nearly nothing has been firmly established. Therefore, the following paragraphs are, to a large extent, speculative and should be read with more than average criticism. The relations to be discussed are listed in Figure 5.

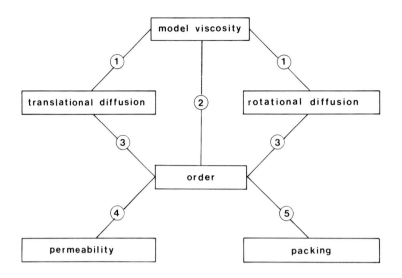

FIGURE 5.   Possible relations between fluidity parameters.

A. Diffusion — Model Viscosity

For a spherical particle in a slow viscous motion through a three-dimensional isotropic fluid, the mobilities of Equation 2 are proportional to the fluidity;

$$b_T^{-1} = 6\pi\alpha\eta; \qquad b_R^{-1} = 8\pi\alpha^3\eta \tag{8}$$

where α denotes the particle radius and η is the viscosity. The ratio between these two mobilities is independent of the viscosity,

$$b_T/b_R = \frac{4}{3}\alpha^2 \tag{9}$$

A hydrodynamic model for a membrane has been treated by Saffman and Delbrück.[9] It is depicted in Figure 6. A protein (represented by a cylinder in Figure 6) is embedded in a lipid bilayer membrane bounded by aqueous phases on both sides. The particle is restricted to move laterally in the x-y plane ($<z^2> = 0$), and to rotate around the z-axis ($<\theta^2> = 0$). In the membrane, the drag friction in the plane is determined by the viscosity $\eta_\perp$ and along the normal by $\eta_\parallel$. The viscosity of the water bounding is η'. For a restricted motion of the protein only $\eta_\perp$ and η' are relevant. It is supposed that $\eta' \ll \eta_\perp$. The rotational mobility reads:[9]

$$b_R^{-1} = 4\pi\alpha^2 d\eta_\perp \tag{10}$$

where d denotes the thickness of the membrane sheet, which has been put equal to the height of the cylinder, and α now stands for the radius of the cylindrical "protein". The translational mobility cannot be derived without referring to the viscosity of the water, for otherwise Stokes' paradox is hard to circumvent.[9] (Stokes' paradox demonstrates that there is no solution for a slow viscous flow in two-dimensions.[8,9]) Saffman and Delbrück find,[9]

$$b_T/b_R = \alpha^2\left(\log\frac{\eta_\perp d}{\eta'\alpha} - \gamma\right) \tag{11}$$

where γ is Euler's constant, $\gamma = 0.5772$.[9] This gives an order of magnitude for $D_T/D_R = b_T/b_R$ of about 10^{-13}cm^2 for a typical membrane protein in agreement with the experiment.[9]

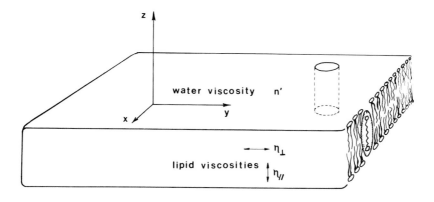

FIGURE 6.   The hydrodynamic model (see text).

B. Order — Model Viscosity

The lipid order and dynamics are often indirectly measured using fluorescence or spin probes[3-6,13,21-25] which can be, at least in many cases, considered as disc–like or rod–like objects exhibiting thermal rotation in the lipid bilayer. For the description of this motion, two different models can be found in the literature:

1.   A molecular model for rotational diffusion in an ordering potential; the relevant parameters are the diffusion constants $D_{\parallel}$, $D_{\perp}$, and the order parameter S.[10]
2.   A hydrodynamic model for the rotation of the probe in an anisotropic fluid with cylindrical symmetry around the normal. (See also Figure 6, where a probe is represented as a cylindrical object between the hydrocarbon tails.) Here the viscosity tensor has two components $\eta_{\perp}$ and $\eta_{\parallel}$ for resistance to motion parallel and perpendicular to the symmetry axis, respectively, with $\eta_{\parallel} < \eta_{\perp}$;[17] another relevant parameter is the viscosity anisotropy R, defined as:

$$R = (\eta_{\parallel} - \eta_{\perp})/(\eta_{\parallel} + 2\eta_{\perp}) \qquad (12)$$

These models are complementary and it should be possible to translate results calculated for the first into the language of the second and vice versa; for example, consider the relation

$$S = -2R \qquad (13)$$

According to Relation 13, a particle rotating in a gradient of viscosity will spend more time in regions where the viscosity is smaller. That means that it assumes orientation preferably along the membrane normal which can be described in terms of an order parameter S. This relation is in agreement with the limiting cases of $\eta_{\parallel} \ll \eta_{\perp}$ where S must be unity and of $\eta_{\parallel} = \eta_{\perp}$ where S must equal zero.[24]

C. Diffusion — Order

It has been observed that the wobbling diffusion constant of the fluidity probe diphenylhexatriene (DPH) is in a first approximation inversely proportional to the order parameter.[23] The hindered wobbling diffusion of DPH is mediated by slowing the probe down through a viscous drag (this reflects the effect of $D_{\perp}$ on the steady-state fluorescence anisotropy, r_s, of the probe), and by restricting the range of the rotation (this corresponds to the effect of S on r_s).

Thus r_s depends both on S and $D_{\perp}$. For DPH a semi-empirical relation has been derived that can be helpful in separating both contributions.[23-25] It has been shown that, if r_s is large (larger than about 0.2), it is dominated by the contribution of the orientational order, whereas in membranes with lower r_s the diffusion has the most dominant contribution. Therefore, S can be readily estimated if r_s is high and $D_\perp$ can be obtained if r_s is small.[23-25] The time-resolved method allows the resolution of both contributions within one experiment.[21,22] In addition, this technique provides more detailed information on the orientational order.[26]

D. Order — Permeability

The molecular mechanism for permeability could be coupled to the rapid diffusion of kinks in the hydrocarbon chains across the membrane.[27] A kink would propagate across a lipid bilayer in the order of 10^{-6} sec and may thus be an effective tunnel for a small molecule.[6] Although kink mobility might well account for the relatively high water permeability through fluid lipid bilayers,[27] no support for the role of kinks has been found in computer simulations of lipid membranes.[28] Other fluctuations in chain configurations probably facilitate the transport of small molecules across the bilayer.

E. Order — Packing

Figure 3 indicates a correlation between orientational order and packing, i.e., the number of molecules per unit area. If the order is high, the hydrocarbon chains are extended and allow neighboring lipids to approach each other closely; on the other hand, if the orientational order is low, the chains have irregular conformation and thus occupy a larger volume or area; therefore, A decreases with increasing S, where A is the area per lipid. The relation order—packing can be quantified as follows: assuming that volume changes associated with order variations are negligible to a first approximation, it holds that the packing is proportional to the relative chain length z, defined as the effective length of a lipid chain l divided by its length in the extended conformation l_0:

$$z = A_0/A = l/l_0 \tag{14}$$

where A_0 is the minimal area per lipid. Data collected from the literature which combine S and z are presented in Figure 7 and Tables 1a and 1b. The experimental trend shown in Figure 7 can be described by the approximate relation

$$z^3 < S < z^2 \tag{15}$$

It is interesting that the theoretical results of Marcelja[32] are in agreement with Equation 15, in lateral pressures π of 50 or 20 dyn/cm, which seem to be realistic values.[33]

Apparently, packing is a good parameter to represent the lipid disorder. This conclusion has also been reached by Mely et al.,[29] who have shown that the order parameter profiles for two lamellar systems with the same area per polar head, but with different water contents and temperatures, are virtually identical.[29]

Relation 15 allows the estimation of the surface pressure in a bilayer by comparing order parameters in bilayers with packing data in monolayers. Using relation 15 introduced above, the A (Π,T) data can be translated into S(Π,T) data. It is assumed that a monolayer is equivalent to a bilayer provided that the surface pressure is adjusted.[33] Comparison of monolayer S(Π,T) data with bilayer S(T) data allows us, therefore, to estimate the surface pressure in a bilayer. This is shown in Figure 8, which indicates that the surface pressure is high in the rigid (high order) phase, changes sharply at the phase transition, and is low in the fluid (low order) phase. Fulford and Peel[35] have proposed an estimation of the lateral pressure from fluorescence anisotropy measurements.[35] Although their result deviates considerably from Figure 8 in the rigid phase, there is agreement in the fluid phase and the trend predicted is the same.

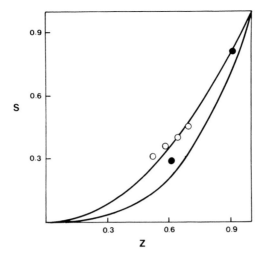

FIGURE 7. The relation between the order parameter S and the relative chain length z. Data, listed in Table 1, are shown for potassium laurate (o, from Reference 29) and for dipalmitoylphosphatidylcholine (●, from References 30 and 31). The upper line describes $S = z^2$ and the lower line describes $S = z^3$.

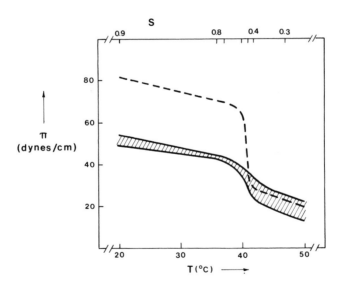

FIGURE 8. Range of lateral pressure values as a function of temperature and order parameter for a bilayer of dipalmitoylphosphatidylcholine. The dotted line is the estimation proposed by Fulford and Peel.[35] The solid curves are obtained by applying Equation 15 to monolayer data from Reference 34 and putting the resulting S equal to values given in Reference 31.

Table 1A
ORDER AND PACKING DATA
FOR POTASSIUM LAURATE IN
WATER

T (°C)	W	A (Å²)	z	S
110	30%	41	0.52	0.31
51	30%	36	0.58	0.36
50	24%	32.8	0.64	0.41
31	21%	30.7	0.68	0.45

Note: Data are obtained from Reference 29. T is the temperature and W is the water content (weight %). z is calculated from $z = A_0/A$ with $A_0 = 21$ Å². S is the segmental order parameter averaged over the chain, obtained from deuterium magnetic resonance. The approximate relation $S = 2|S_{CD}|$ has been used, where S_{CD} denotes the order parameter of a CD bond.

Table 1B
ORDER AND PACKING DATA
FOR DIPALMITOYL-
PHOSPHATIDYLCHOLINE IN
WATER

T (°C)	W	d (Å)	z	Sª
35	5%	48.3	0.91	0.81
49	5%	36.7	0.61	0.29

Note: S is obtained from Reference 31 and z from Reference 30; z is calculated from $z = d_c/d_{co}$, where d_c is the thickness of the hydrocarbon part of the membrane, d_{co} is the maximal d_c, which is taken as $d - 13$ Å; the membrane thickness d is measured with X-ray diffraction techniques.[30] T is the temperature and W is the water content (weight %).

[a] S data are obtained from fluorescence anisotropy data using diphenylhexatriene as a probe. It is assumed here that this order parameter equals the averaged chain order parameter. Comparison of deuterium magnetic resonance data with fluorescence anisotropy data show that this is indeed a reasonable approximation (maximal deviation 10%).[30,31]

Lateral pressure in membranes can be measured by studying the action of a group of phospholipases.[36] In the case of intact erythrocytes, phospholipases purified from various sources were divided into two groups. One group of enzymes (group 1) could not hydrolyze the phospholipids of intact erythrocyte membranes, whereas the other group (group 2) could. The enzymes of group 1 were unable to degrate monolayers at surface pressures above 33 dyn/cm, while those of group 2 could attack monomolecular films at all surface pressures. From this, one can conclude that the lipid packing in the erythrocyte membrane is comparable with a lateral pressure of about 33 dyn/cm.[4,36]

It is also interesting to compare the effect of cholesterol in monolayers and bilayers. In monolayers, cholesterol acts as a ''condenser''; it reduces the average molecular area if the lipids are in the fluid state.[37] In bilayers, cholesterol increases the order parameter in the fluid phase.[21,38] This is in agreement with its effect in monomolecular films, because the order parameter is believed to reflect the packing in the membrane.[25,35,39] However, if the bilayer is in the rigid state, i.e., below the chain melting, cholesterol tends to fluidize the membrane. For example, if cholesterol is introduced in the fluid phase of a dipalmitoyl-phosphatidylcholine bilayer, the water permeability is strongly reduced, whereas cholesterol increases the permeability in the low-temperature phase. In the presence of 33 mol% cholesterol, the abrupt phase transition, which was clearly visible in the pure lipid-water system, is completely abolished.[49] In biological membranes at ambient temperatures, cholesterol thus imposes on the heterogeneous population of phospholipids a condition of ''intermediate fluidity'',[50] and causes a smoothening or disappearance of phase transitions. In general, the overall effect of cholesterol in these membranes is rigidification.[25]

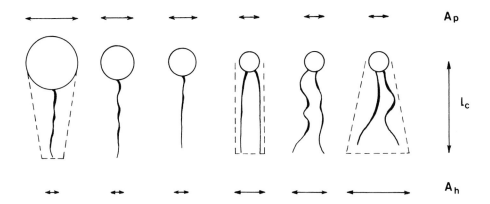

FIGURE 9. Schematic diagram of lipid shapes, varying from a truncated cone via a cylinder to an inverted truncated cone. The chain length l_c, the area per polar head A_p, and the hydrocarbon area of an average cross section A_c are indicated. More realistic pictures of lipids can be found in References 30 and 41.

IV. MOLECULAR SHAPE AND FORMS OF ORGANIZATION

Parameters for the molecular shape are the chain length l_c, the area per polar headgroup A_p, and the cross section of the hydrocarbon part A_h. Other relevant parameters are the polarity of the headgroup and the hydrophobicity of hydrocarbon tails. Variations in the lipid shape are schematically depicted in Figure 9.

A. Polymorphism in Lipid-Water Systems

Lipid molecules in water aggregate into a variety of structures, which can be classified in more than 18 different phases based on symmetry.[2,42-45] The phases differ in translational and rotational order. Translational order refers to periodicity; the system can be periodic in three dimensions (crystalline phases), in one or two dimensions (mesomorphic phases), or totally aperiodic. Rotational order may refer to the orientations of molecules or to those of elongated aggregates of lipids. The biologically relevant phases are listed below.

I_{hp} The isotropic phase of a molecular solution of lipids in water. The concentration of the lipid is below the critical micelle concentration.

M_{hp} The micellar phase where aggregates of lipids are dissolved in water (Figure 10A). The concentration of lipid molecules is above the critical micelle concentration. The micelle is rotationally and translationally disordered.

H_{hp} The hexagonal type I phase, often denoted as H_I. This phase is a two-dimensional hexagonal array of rods with polar surfaces and hydrocarbon interiors (Figure 10B). The ratio polar area/hydrocarbon area is larger than in the M_{hp} phase, but smaller than unity.

L   The lamellar phase, in which the polar groups are localized in planes facing the water. The lamellae have a bilayer structure and can be planar or curved. The area per polar head is close to the hydrocarbon area of a molecule. The L phase can be subclassified according to chain order:

Lα   In this phase the hydrocarbon chains in the bilayer interior are disordered (Figure 11A).

Lβ   The Lβ phase exhibits a bilayer structure where the chains are rigid and parallel (Figure 11B).

Lβ'   This phase is similar to Lβ, except that the chains are tilted normal to the bilayer (Figure 11C). The tilt angle varies with the water content.

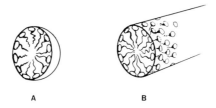

FIGURE 10.   A micelle in the M_{hp} phase (A) and a rod in the H_{hp} phase (B).

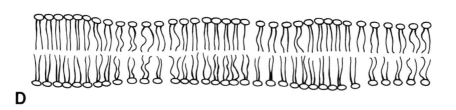

FIGURE 11.   Bilayers in the phases Lα (A), Lβ (B), Lβ′ (C), and Lαβ (D).

Lαβ   The Lαβ phase represents a structure intermediate between Lα and Lβ, as depicted in Figure 11D. It occurs in the α-β transition, which is a gradual transition, if more than one lipid component is present. The components tend to separate into different domains with different compositions. However, there is a striking difference to ordinary three-dimensional systems, where lateral phase separation usually leads to a complete separation of the components.[6] In Lαβ a mosaic of fluid and lipid domains is found, generally having different curvatures.[6] The boundaries are not static, but are at a constant motion.

H_{ph} The hexagonal phase type II, frequently denoted as H_{II}. Here cylinders of water are embedded in lipid. The area per polar group is smaller than the average cross section in the hydrocarbon region (Figure 12A).

M_{ph} The micellar phase consisting of aggregates in lipid (Figure 12B). The ratio polar area/ hydrocarbon area is smaller than in H_{ph}.

I_{ph} A lipid dispersion where the lipid phase contains a very low concentration of water.

B. Phase Transitions

In a stable lipid aggregate the packing will be such that the area per polar head A_p' at the interface is larger than the area of the polar head itself, A_p, and the area per hydrophobic part A_h' will be close to the A_h of the hydrocarbon tails. The lamellar structure of a lipid bilayer may become unstable at high lipid concentrations giving way to the inverted hexagonal

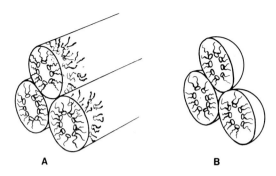

FIGURE 12.   Structure of the H_{ph} (A) and the M_{ph} phase (B).

H_{ph} phase and further to inverted spherical micelles (M_{ph}). It is clear that different lipids have a different tendency to form one phase or another. Dynamic packing properties determine the favored structure at a given temperature and water content. For example, dipalmitoylphosphatidylcholine at T = 35°C and W = 5% (W is water concentration by weight) is in the Lβ phase. But when W is increased, the distance between the polar heads increases also, while the chains keep their original packing. As a result, a transition to the Lβ′ phase occurs and at T = 35°C and W = 50% the chains are tilted with respect to the bilayer normal by an angle of 27°.[30] Variations in distances between polar heads can also be induced by changes in salt concentrations or pH.[46] Similarly, the addition of salt to a micellar solution reduces A_p' and leads to larger micelles due to a partial screening of the electrostatic head group repulsion.[47] Charge screening is, therefore, equivalent to reducing the size of the polar groups. The effect of adding water is to increase the area per polar head at the interface.

Temperature alterations affect the hydrocarbon tails. As discussed above, when the temperature increases, the chain order decreases, associated with a growing number of *gauche* isomers. As a result, the area per chain becomes larger. At the so-called chain melting transition (see before) the order suddenly changes and the system transforms from a Lα into a Lβ phase or vice versa. This phase change has been extensively discussed by others.[3,6,18,32] In lipid-water systems containing one lipid component, the α-β transition is sharp, whereas in multicomponent systems and in biological membranes the chain melting occurs as a broad transition.[3,18,43,48] It should be noted that various kinds of curved lamellar structures are possible. Examples are multilamellar or unilamellar vesicles and of course biological membranes, which are curved to some extent. Other possibilities could be cylindrical bilayers or bicontinuous lamellae resembling a structure consisting of interconnected pipes and spheres.

Proteins are amphiphilic as well. Therefore, protein-lipid-water systems are expected to behave similarly. However, proteins are more rigid and cannot so easily adjust themselves to the environment as flexible lipids do. This is probably the reason why proteins and steroids tend to segregate into α regions of the Lαβ phase. As far as *structure* is concerned, the distinction between lipids and proteins is artificial. A classification into rigid molecules (e.g., steroids, proteins) and flexible molecules (e.g., lipids above T_M, unfolded proteins) is more useful.[47] The empirical rules (1) A_p' becomes larger with increasing W, and (2) A_h' becomes larger with increasing T allow for a qualitative understanding of phase diagrams, i.e., plots of T vs. W at a constant lipid concentration in which the stability regions of the various phases are indicated.

Phase transitions may take part in physiological processes. The lung surfactant, for example, exhibits a phase transition near 37°C. The system can, therefore, adjust the interface area upon pressure variations in order to allow mechanical support and adequate gas transport.[56,57]

Another example concerns the Lαβ phase, where the membrane is in the middle of the α-β transition. A functional advantage of this phase is a high lateral compressibility. As a result, the membrane is able to maintain a constant lateral chain pressure when the membrane area is changed. This may defend the membrane against thermal or osmotic shock, and against area changes due to incorporation or removal of membrane material.[47,53] Lateral displacements in the Lαβ phase are also relevant immunologically, since segregation modulates the immunological recognition.[54] This organization in the plane may affect the boundary lipids, which regulate the function of membrane proteins.[18,55]

Nonbilayer structures can also be of biological significance. Locally the membrane may fluctuate into a hexagonal phase or inverted micellar structure, if shape and hydration conditions allow this. Transient lipid particles may then exist in the membrane, which probably play a role in transport across the bilayer.[58-60] Furthermore, these structures,[61] or point defects,[62] could facilitate membrane fusion.

V. THE PROTEIN POSITION

Membrane-bound proteins can be divided into integral and peripheral proteins.[1] Integral proteins are intercalated into the hydrophobic core of the lipid bilayer, while the hydrophobic portion of the molecules protrude into the inside and/or outside aqueous phase. Peripheral proteins are only weakly bound to the internal or external surface. Integral proteins can be subdivided into cross-membrane, with hydrophobic parts on both sides, and integral proteins that protrude only into one aqueous phase.

The position of a membrane protein is under thermodynamic control.[63] Hydrophilic forces try to pull it into the water, while hydrophobic forces operate in the opposite direction. Together they determine the equilibrium position, which can in principle be changed by membrane modification, as has been noted by Borochov and Shinitzky, who have formulated the "vertical displacement hypothesis".[64] This says that upon increase of membrane rigidity, proteins will be vertically displaced to a position where they will be more exposed to the aqueous domain on either side of the membrane. Evidence for this idea has been obtained using spectroscopic techniques,[64,65] chemical methods,[66] and ligand binding.[65,67-69] According to this hypothesis, the degree of exposure of receptors, antigens, and other functional proteins can be modified by changing the lipid state both in vivo and in vitro.

A. Analogy to the Law of Archimedes

According to the law of Archimedes, a floating body in a three-dimensional fluid can be vertically displaced by changing the density of the fluid (by adding salt, for instance). This is analogous to the problem of a protein partly immersed in a two-dimensional fluid where the packing can be changed. The hydrostatic pressure of the three-dimensional case is analogous to the lateral pressure in two dimensions and the density is analogous to the packing. A protein can be displaced if it is asymmetrically shaped. It is possible to estimate the shift in location, if it is assumed that the free energy contribution from the binding to the membrane varies quadratically with the displacement and the repulsive contribution from the lateral pressure varies linearly, as a result of a linear dependence in the average cross section of the protein. This model is depicted in Figure 13. Consider first a typical cross-membrane protein-like glycophorin. Bell has estimated the free energy change as one glycophorin molecule is pulled from the membrane into the water: 1 eV to transfer the hydrophobic core from the lipid into the water and, in addition, about 1.6 eV to bring the cytoplasmic tail through the lipid.[73] Assuming an increase in the lateral pressure of 40 dyn/cm = 2.5×10^{-3} eV/Å^2 upon rigidification, a difference in top and bottom radius of 10 Å and a thickness of 40 Å for the lipophilic part of the protein correspond to a vertical displacement of about 3 Å. Realizing that a lateral pressure change of 40 dyn/cm is rather

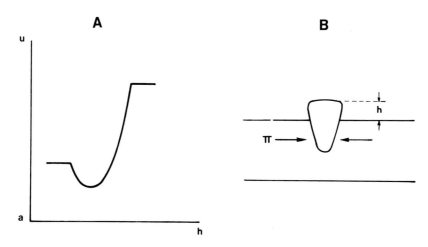

FIGURE 13. The free energy u as a function of the protrusion h (A); the lateral pressure II acting on a membrane protein that protrudes to a distance h above the membrane surface (B).

high, this result indicates that the vertical shift for a cross-membrane protein is negligible, at least in the context of the present model. The situation for an integral protein protruding only at one side is different, since the binding is weaker and the barrier due to the cytoplasmic tail is absent. Taking the same values as above with omission of this barrier, the vertical displacement becomes about 8 Å. In this static picture, oscillations about the equilibrium position are ignored. Such fluctuations will have much larger amplitudes in the shifted location, because there the binding to the membrane is weaker. We may conclude that, for an integral protein that does not traverse the whole membrane, a vertical displacement is to be expected in the present model. On the other hand, a positional shift of a cross-membrane protein is unlikely, the more so as the membrane becomes also thicker upon rigidification. Indeed, the expression of such a protein may diminish as a result of an increase in membrane thickness.

B. The Gerson Model

Gerson has proposed a quantitative model in which the vertical position of a membrane protein is derived from interfacial energies.[74,75] His model is schematically depicted in Figure 14. It consists of three isotropic homogeneous phases: the membrane, the protein, and the water. Interfaces between immiscible phases have an associated interfacial free energy (γ, energy per area). There are three interfacial free energies here: between the membrane and the water, γ_{mw}; between the protein and the water, γ_{pw}; and between the membrane and the protein, γ_{pm}. The membrane is taken as a planar hydrophobic bulk phase, which is thick relative to the protein. The protein is modeled as a sphere of a radius r. The distance between top and surface is h. The total interfacial free energy reads:[80]

$$U = U_0 - 2(\gamma_{mw} - \gamma_{pw} + \gamma_{pm})\,\pi hr + \gamma_{mw}\pi h^2 \qquad (16)$$

where U_0 is the energy, when the protein is fully submerged in the membrane. The equilibrium position corresponds to the minimum of U: if $\gamma_{pm} > \gamma_{pw} + \gamma_{mw}$, the protein will be completely detached from the membrane and will dissolve in the water, but if $\gamma_{pm} < \gamma_{pw} - \gamma_{mw}$, the protein will be completely submerged in the membrane. In the intermediate region, $\gamma_{pw} - \gamma_{mw} < \gamma_{pm} < \gamma_{pw} + \gamma_{mw}$, the vertical placement is given by

$$h/r = 1 + (\gamma_{pm} - \gamma_{pw})/\gamma_{mw} \qquad (17)$$

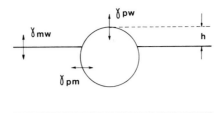

FIGURE 14.   The Gerson model for the positioning of a spherical protein particle at the membrane surface. Interfacial free energies, γ, at a protein lipid interface.

The magnitude of γ_{pm} may be estimated from γ_{pw} and γ_{mw} using an empirical relation.[74,75] In this model, addition of cholesterol makes the membrane more hydrophobic. As a result, the relatively hydrophilic protein would be expected to rise out of the membrane to some extent.[74,75]

C. Curvature Changes

Another possibility is that the curvature of the bilayer changes near the protein upon membrane modification. This could be due to the onset of a phase transition into a nonbilayer structure, or it could be a result of a phase separation, where the domains differ in curvature. Indeed, there is some evidence from freeze-fracture electron microscopy that vertical displacements are only seen if lateral phase separation occurs.[76-78] In this respect, cholesterol can induce phase separation.[79] The combined occurrence of vertical and lateral displacements has been recently confirmed using low angle X-ray diffraction by Funk et al.[48] These authors conclude that the temperature-induced mass shift observed results from a vertical displacement of proteins and that this is triggered by an abrupt lateral redistribution of lipids in the Lαβ phase.

Curvature changes due to instability of the bilayer could be visualized as waves on the sea of lipids in the Fluid Mosaic model. A vertical displacement is then a dynamic one: the oscillated protein is more exposed on the average.

VI. CONCLUSION

"Membrane fluidity" is a property of the lipids. It is often used in a loose context to describe how disordered the hydrocarbon chains are (structural information) and/or how easily they move (dynamical information). In a quantitative formulation one physical parameter is not sufficient to characterize both structure and dynamics of the lipids. Membrane fluidity rather represents a set of parameters for the packing, the rotational and translational diffusion, the permeability, and the order. These fluidity parameters are related in a way that is not yet fully understood. They describe the physical state of the lipids, which have a prominent effect on the physiological functions of the membrane. There are now experimental indications that the role of lipids is much more important than outlined in this chapter. It is appropriate to conclude with the words of Dr. Luzzati: "Lipids are very clever, far more than I am."[81]

ACKNOWLEDGMENTS

Thanks are due to Henk Broxterman and Ernst de Bruijn for a critical reading of the manuscript and to Bela Mulder for stimulating discussions. Financial support from the Koningin Wilhelmina Fonds, Netherlands Cancer Foundation is gratefully acknowledged.

REFERENCES

1. **Singer, S. J. and Nicolson, G. L.,** The fluid mosaic model of the structure of cell membranes, *Science,* 175, 720, 1972.
2. **Luzzati, V.,** X-ray diffraction studies of lipid-water systems, in *Biological Membrane,* Chapman, D., Ed., Academic Press, New York, 1968, chap. 3.
3. **Nicolson, G. L., Poste, G., and Ji, T. H.,** The dynamics of cell membrane organization, in *Dynamic Aspects of Cell Surface Organization,* Poste, G. and Nicolson, G. L., Eds., North-Holland, Amsterdam, 1981, chap. 1.
4. **Seelig, J.,** Physical properties of model membranes and biological membranes, in *Membranes and Intercellular Communication,* Balian, R., Chabre, M., and Devaux, F., Eds., North-Holland, Amsterdam, 1981, 18.
5. **Cherry, R. J.,** Rotational and lateral diffusion of membrane proteins, *Biochim. Biophys. Acta,* 559, 289, 1979.
6. **Sackmann, E.,** Dynamic molecular organization in vesicles and membranes, *Ber. Bunsenges. Phys. Chem.,* 82, 981, 1978.
7. **Shinitzky, M. and Henkart, P.,** Fluidity of cell membranes — current concepts and trends, *Int. Rev. Cytol.,* 60, 121, 1979.
8. **Landau, L. D. and Lifshitz, E. M.,** *Fluid Mechanics,* Pergamon Press, New York, 1959.
9. **Saffman, P. G. and Delbrück, M.,** Brownian motion in biological membranes, *Proc. Natl. Acad. Sci. U.S.A.,* 72, 3111, 1975.
10. **Zannoni, C., Arcioni, A., and Cavatorta, P.,** Fluorescence depolarization in liquid crystals and membrane bilayers, *Chem. Phys. Lipids,* 32, 179, 1983.
11. **Kooyman, R. P. H., Levine, Y. K., and Van der Meer, W.,** Measurement of second and fourth rank order parameters by fluorescence polarization experiments in a lipid membrane system, *Chem. Phys.,* 60, 317, 1981.
12. **Van der Meer, W., Kooyman, R. P. H., and Levine, Y. K.,** A theory of fluorescence depolarization in macroscopically ordered membrane systems, *Chem. Phys.,* 66, 39, 1982.
13. **Shinitzky, M. and Barenholz, Y.,** Fluidity parameters of lipid regions determined by fluorescence polarization, *Biochim. Biophys. Acta,* 515, 367, 1978.
14. **Kornberg, R. D. and McConnell, H. M.,** Inside-outside transitions of phospholipids in vesicle membranes, *Biochemistry,* 10, 1111, 1971.
15. **Rothman, J. E. and Lenard, J.,** Membrane asymmetry, *Science,* 195, 743, 1977.
16. **Batschinski, A. J.,** Untersuchungen über die innere Reibung der Flussigkeiten, *Z. Physik. Chem.,* 89, 643, 1913.
17. **Shinitzky, M. and Yuli, I.,** Lipid fluidity at the submacroscopic level; determination by fluorescence polarization, *Chem. Phys. Lipids,* 30, 261, 1982.
18. **Kimelberg, H. K.,** Membrane fluidity and the activity of membrane-bound enzymes, NATO Advanced Study Institute, Physical Methods on Biological Membranes, Altavilla Milicia, Italy, September 20 to October 2, 1982.
19. **Fahey, P. F. and Webb, W. W.,** Lateral diffusion in phospholipid bilayer membranes and multilamellar liquid crystals, *Biochemistry,* 17, 3046, 1978.
20. **Van Zoelen, E. J. J., Hendriques de Jesus, C., De Jonge, E., Mulder, M., Blok, M. C., and De Gier, J.,** Nonelectrolyte permeability as a tool for studying membrane fluidity, *Biochim. Biophys. Acta,* 511, 335, 1978.
21. **Heyn, M. P.,** Determination of lipid order parameters and rotational correlation times from fluorescence depolarization experiments, *FEBS Lett.,* 108, 359, 1979.
22. **Jähnig, F.,** Structural order of lipids and proteins in membranes: evaluation of fluorescence anisotropy data, *Proc. Natl. Acad. Sci. U.S.A.,* 76, 6361, 1979.
23. **Pottel, H., Van der Meer, W., and Herreman, W.,** Correlation between the order parameter and the steady-state fluorescence anisotropy and an evaluation of membrane fluidity, *Biochim. Biophys. Acta,* 730, 181, 1983.
24. **Van der Meer, W., Pottel, H., and Herreman, W.,** Correlation between steady-state and time-resolved fluorescence anisotropy, NATO Advanced Study Institute, Physical Methods on Biological Membranes, Altavilla Milicia, Italy, September 20 to October 2, 1982.
25. **Van Blitterswijk, W. J., Van Hoeven, R. P., and Van der Meer, W.,** Lipid structural order parameters (reciprocal of fluidity) in biomembranes derived from steady-state polarization measurements, *Biochim. Biophys. Acta,* 644, 323, 1981.
26. **Ameloot, M., Van der Meer, W., Pottel, H., Herreman, W., Hendrickx, H., and Schröder, H.,** Non-a-priori approach to the analysis of time-resolved fluorescence anisotropy in membranes, to be published.

27. **Träuble, H.**, The movement of molecules across lipid membranes: a molecular theory, *J. Membr. Biol.*, 4, 193, 1971.

28. **Van der Ploeg, P. and Berendsen, H. J. C.**, Molecular dynamics simulation of a bilayer membrane, *J. Chem. Phys.*, 76, 3271, 1982.

29. **Mely, B., Charvolin, J., and Keller, P.**, Disorder of lipid chains as a function of their lateral packing in lyotropic liquid crystals, *Chem. Phys. Lipids*, 15, 161, 1975.

30. **Seelig, A. and Seelig, J.**, The dynamic structure of fatty acyl chains in a phospholipid bilayer measured by deuterium magnetic resonance, *Biochemistry*, 13, 4893, 1974.

31. **Lakowicz, J. R., Prendergast, F. G., and Hogen, D.**, Differential polarized phase fluorometric investigations of diphenylhexatriene in lipid bilayers. Quantitation of hindered depolarizing rotations, *Biochemistry*, 18, 508, 1979.

32. **Marcelja, S.**, Chain order in liquid crystals. II. Structure of bilayer membranes, *Biochim. Biophys. Acta*, 367, 165, 1974.

33. **Meraldi, J. P. and Schlitter, J.**, A statistical mechanical treatment of fatty acyl chain order in phospholipid bilayers and correlation with experimental data, *Biochim. Biophys. Acta*, 645, 193, 1981.

34. **Philips, M. C. and Chapman, D.**, Monolayer characteristics of saturated 1,2-diacyl phosphatidylcholines (lecithins) and phosphatidylethanolamines at the air-water interface, *Biochim. Biophys. Acta*, 163, 301, 1968.

35. **Fulford, A. J. C. and Peel, W. E.**, Lateral pressures in biomembranes estimated from the dynamics of fluorescent probes, *Biochim. Biophys. Acta*, 598, 237, 1980.

36. **Demel, R. A., Geuris van Kessel, W. S. M., Zwaal, R. F. A., Roelofsen, B., and Van Deenen, L. L. M.**, Relation between various phospholipase actions on human red cell membranes and the interfacial phospholipid pressure in monolayers, *Biochim. Biophys. Acta*, 406, 97, 1975.

37. **Demel, R. A., Van Deenen, L. L. M., and Pethica, B. A.**, Monolayer interactions of phospholipids and cholesterol, *Biochim. Biophys. Acta*, 135, 11, 1967.

38. **Kawato, S., Kinosita, K., Jr., and Ikegami, A.**, Effect of cholesterol on the molecular motion in the hydrocarbon region of lecithin bilayers studied by nanosecond fluorescence techniques, *Biochemistry*, 17, 5026, 1978.

39. **Kleinfeld, A. M., Dragsten, P., Klausner, R. D., Pjura, W. J., and Matayoshi, E. D.**, The lack of relationship between fluorescence polarization and lateral diffusion in biological membranes, *Biochim. Biophys. Acta*, 649, 471, 1981.

40. **Shinitzky, M.**, Membrane fluidity and receptor functions, *Biomembranes*, 12, 1983.

41. **Lehninger, A. L.**, *Biochemistry*, Worth Publishers, New York, 1970.

42. **Tardieu, A., Luzzati, V., and Reman, F. C.**, Structure and polymorphism of the hydrocarbon chains of lipids: a study of lecithin-water phases, *J. Mol. Biol.*, 75, 711, 1973.

43. **Ranck, J. L., Mateu, L., Sadler, D. M., Tardieu, A., Gulik-Krzywicki, T., and Luzzati, V.**, Order-disorder conformational transitions of the hydrocarbon chains of lipids, *J. Mol. Biol.*, 85, 249, 1974.

44. **Brown, G. H. and Wolken, J. J**, *Liquid Crystals and Biological Structures*, Academic Press, New York, 1979.

45. **Winsor, P. A.**, Binary and multicomponent solutions of amphiphilic compounds, *Chem. Rev.*, 68, 1, 1968.

46. **Jähnig, F., Harlos, K., Vogel, H., and Eibl, H.**, Electrostatic interactions at charged lipid membranes. Electrostatically induced tilt, *Biochemistry*, 18, 1459, 1979.

47. **Israelachvili, J. N., Marcelja, S., and Horn, R. G.**, Physical principles of membrane organization, *Q. Rev. Biophys.*, 13, 121, 1980.

48. **Funk, J., Wunderlich, F., and Kreutz, W.**, Temperature-induced vertical shift of proteins in membranes, *J. Mol. Biol.*, 161, 561, 1982.

49. **De Gier, J., Blok, M. C., Van Dijck, P. W. M., Momber, C., Verkleij, A. J., Van der Neut-Kok, E., and Van Deenen, L. L. M.**, Relations between liposomes and biomembranes, *Ann. N. Y. Acad. Sci.*, 308, 399, 1978.

50. **Oldfield, E. and Chapman, D.**, Dynamics of lipids in membranes: heterogeneity and the role of cholesterol, *FEBS Lett.*, 23, 285, 1972.

51. **Cullis, P. R. and De Kruiff, B.**, Polymorphic phase behaviour of lipid mixtures as detected by ^{31}P NMR. Evidence that cholesterol may destabilize bilayer structure in membrane systems containing phosphatidylethanolamine, *Biochim. Biophys. Acta*, 507, 207, 1978.

52. **Cullis, P. R. and De Kruiff, B.**, Lipid polymorphism and the functional roles of lipids in biological membranes, *Biochim. Biophys. Acta*, 559, 399, 1978.

53. **Shimshick, E. J. and McConnell, H. M.**, Lateral phase separation in phospholipid membranes, *Biochemistry*, 12, 2351, 1973.

54. **Ruysschaert, J. M., Tenenbaum, A., and Berliner, C.**, Correlation between lateral lipid phase separation and immunological recognition in sensitized liposomes, *FEBS Lett.*, 81, 406, 1977.

55. **Sandermann, H.**, Regulation of membrane enzymes by lipids, *Biochim. Biophys. Acta*, 515, 209, 1978.

56. **Träuble, H., Eibl, H., and Swada, H.,** Respiration — a critical phenomenon? Lipid phase transition in the lung alveolar system, *Naturwissenschaften,* 61, 344, 1974.
57. **Bangham, A. D., Morley, C. J., and Philips, M. C.,** The physical properties of an effective lung surfactant, *Biochim. Biophys. Acta,* 573, 552, 1979.
58. **De Kruiff, B., Verkleij, A. J., Van Echteld, C. J. A., Gerritsen, W. J., Mombers, C., Noordam, P. C., and De Gier, J.,** The occurrence of lipidic particles in lipid bilayers as seen by ^{31}P NMR and freeze-fracture electronmicroscopy, *Biochim. Biophys. Acta,* 555, 200, 1979.
59. **Noordam, P. C., Van Echteld, C. J. A., De Kruiff, B., Verkleij, A. J., and De Gier, J.,** Barrier characteristics of membrane model systems containing unsaturated phosphatidylethanolamines, *Chem. Phys. Lipids,* 27, 221, 1980.
60. **Mandersloot, J. G., Gerritsen, W. J., Leunissen-Bijvelt, J., Van Echteld, C. J. A., Noordam, P. C., and De Gier, J.,** Ca^{2+}-induced changes in the barrier properties of cardiolipin/phosphatidylcholine bilayers, *Biochim. Biophys. Acta,* 640, 106, 1981.
61. **Verkleij, A. J., Mombers, C., Gerritsen, W. J., Leunissen-Bijvelt, L., and Cullis, P. R.,** Fusion of phospholipid vesicles in association with the appearance of lipidic particles as visualized by freeze fracturing, *Biochim. Biophys. Acta,* 555, 358, 1979.
62. **Hui, S. W., Stewart, T. P., Boni, L. T., and Yeagle, P. L.,** Membrane fusion through point defects in bilayers, *Science,* 212, 921, 1981.
63. **Tanford, C.,** The hydrophobic effect and the organization of living matter, *Science,* 200, 1012, 1978.
64. **Borochov, H. and Shinitzky, M.,** Vertical displacement of membrane proteins mediated by changes in microviscosity, *Proc. Natl. Acad. Sci. U.S.A.,* 73, 4526, 1976.
65. **Shinitzky, M. and Souroujon, M.,** Passive modulation of blood-group antigens, *Proc. Natl. Acad. Sci. U.S.A.,* 76, 4438, 1979.
66. **Borochov, H., Abbott, R. E., Schachter, D., and Shinitzky, M.,** Modulation of erythrocyte membrane proteins by membrane cholesterol and lipid fluidity, *Biochemistry,* 18, 251, 1979.
67. **Heron, D., Shinitzky, M., Hershkowitz, M., and Samule, D.,** Lipid fluidity markedly modulates the binding of serotonin to mouse brain membranes, *Proc. Natl. Acad. Sci. U.S.A.,* 77, 7463, 1980.
68. **Muller, C. P. and Shinitzky, M.,** Modulation of transferrin receptors in bone marrow by changes in lipid fluidity, *Br. J. Haematol.,* 42, 355, 1979.
69. **Muller, C. P. and Shinitzky, M.,** Passive shedding of erythrocyte antigens induced by membrane rigidification, *Exp. Cell. Res.,* 136, 53, 1981.
70. **Brulet, P. and McConnell, H. M.,** Structural and dynamical aspects of membrane immunochemistry using model membranes, *Biochemistry,* 16, 1209, 1977.
71. **Yasuda, T., Dancey, G. F., and Kinsky, S. C.,** Immunogenicity of liposomal model membranes in mice: dependence on phospholipid composition, *Proc. Natl. Acad. Sci. U.S.A.,* 74, 1234, 1977.
72. **Yuli, I., Wilbrandt, W., and Shinitzky, M.,** Glucose transport through cell membranes of modified lipid fluidity, *Biochemistry,* 20, 4250, 1981.
73. **Bell, G. I.,** Models for the specific adhesion of cells to cells, *Science,* 200, 618, 1978.
74. **Gerson, D. F.,** The biophysics of membrane proteins: vertical displacement, aggregation, patching and cell-cell recognition, in *The Immune System,* Vol. 1, Steinberg, C. M. and Lefkovits, I., Eds., Karger, Basel, 1981, 245.
75. **Gerson, D. F.,** Interfacial free energies and the control of the positioning and aggregation of membrane proteins, *Biophys. J.,* 37, 145, 1982.
76. **Verkleij, A. J., Ververgaert, P. H. J., Van Deenen, L. L. M., and Elbers, P. F.,** Phase transitions of phospholipid bilayers and membranes of *Acheoplasma Laidlawii* B visualized by freeze fracturing electron microscopy, *Biochim. Biophys. Acta,* 288, 326, 1972.
77. **Wunderlich, F., Ronai, A., Speth, V., Seelig, J., and Blumen, A.,** Thermotropic lipid clustering in *Tetrahymena* membranes, *Biochemistry,* 14, 3730, 1975.
78. **Kleeman, W. and McConnell, H. M.,** Lateral phase separations in *E. coli* membranes, *Biochim. Biophys. Acta,* 291, 220, 1974.
79. **Cherry, R. J., Müller, U., Holenstein, G., and Heyn, M. P.,** Lateral segregation of proteins induced by cholesterol in bacteriorhodopsinphospholipid vesicles, *Biochim. Biophys. Acta,* 569, 145, 1980.
80. **Albertson, P. A.,** *Partitioning of Cell Particles and Macromolecules,* John Wiley & Sons, New York, 1971.
81. **Luzzati, V.,** Lipid polymorphism revisited, NATO Advanced Study Institute, Physical Methods on Biological Membranes, Altavilla Milicia, Italy, September 20 to October 2, 1982.

Chapter 3

REGULATION OF CELL MEMBRANE CHOLESTEROL

Richard A. Cooper and Jerome F. Strauss, III

TABLE OF CONTENTS

I. INTRODUCTION

Cholesterol is a component of membranes in all mammalian cells. Surface membranes are rich in cholesterol, whereas mitochondrial membranes possess little cholesterol, and endoplasmic reticulum possesses even less. Among cells of different types, the amount of cholesterol within the surface membrane varies considerably. However, each cell strives to maintain its own peculiar membrane lipid composition, especially the amount of cholesterol relative to the more polar lipids, phospholipids, and glycolipids.

Cells live in vivo in an environment rich in cholesterol, which is carried by plasma lipoproteins. Multiple avenues exist through which this cholesterol may enter cells. Most cells also possess the enzymatic machinery to synthesize cholesterol *de novo*. In normal cells, both the rate of cholesterol accumulation from extracellular pools and the rate of *de novo* sterol synthesis are strictly regulated in order to stabilize the cholesterol content of membranes. This chapter will discuss the mechanisms through which mammalian cells achieve this control and, in particular, how they regulate the amount of cholesterol in their surface membranes. It will also describe some of the consequences for cells when the sterol content of their membranes deviates from normal.

II. CHOLESTEROL IN MEMBRANES

A. Cholesterol-Phospholipid Interactions

The association between polar lipids and sterols is critical in governing the lipid composition of membranes and the distribution of cholesterol among membranes. This is emphasized by the fact that the membranes of all higher organisms, including plants and animals, contain both. In those few instances where sterols are not known to occur in prokaryotes and lower eukaryotes, molecules are present which appear to mimic the sterol structure.[1,2] For example, caratenols are found in certain species of mycoplasma, and tetrahymanol (a triterpene) in *Tetrahymena pyriformis*. Some species of bacteria have also been found to contain triterpenes similar to tetrahymanol. Cholesterol is the predominant sterol in all cells within the animal kingdom, whereas sterols with an alkylated side chain predominate in plants. The structural requirements for sterols in cell membranes are quite specific. They must possess a β-OH on the third carbon, alternating *trans*-antistereochemistry creating a planar ring structure, and an uncyclized side chain at C17 which possesses at least eight carbon atoms.[1,3-5]

The molecular interactions which occur between sterols and polar lipids in membranes are not completely understood. It has been proposed that hydrogen bonding occurs between the carbonyl oxygen of the phospholipid acyl side chains and the 3β-OH of the sterol.[6] Consistent with this view is the fact that neither 3β-thiocholesterol nor the 3α-OH isomer of cholesterol, epicholesterol, interact with phospholipids as efficiently as cholesterol.[4] However, other modifications of either the sterol molecule or the acyl chains of phospholipids that prevent hydrogen bonding do not interfere with sterol-phospholipid interactions,[7,8] and nonpolar interactions may be more critical than hydrogen bonding.

In most biologic membranes, one of the phospholipid acyl chains is saturated and the other has at least one unsaturated *cis* double bond. This asymmetry of acyl chain structure accommodates the asymmetry of the sterol molecule, caused by the protrusion of two angular methyl groups (C18 and C19) from the β plane of the sterol nucleus. Models of sterol-phospholipid interactions suggest that phospholipids are capable of accommodating up to 2 mol of cholesterol per mole of phospholipid. For amounts of sterol in excess of a cholesterol/phospholipid (C/PL) mole ratio of 2.0, some other molecular configuration is required.[9]

The interactions between sterols and phospholipids have a number of important consequences for membrane structure. For example, sterols increase the efficiency of packing of

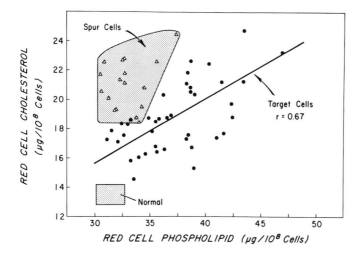

FIGURE 2. Cholesterol and phospholipid content of red cells in patients with liver disease. Most patients have "target" red cells in which membranes are enriched with both cholesterol and phospholipid. A minority of patients have "spur" red cells in which membranes are disproportionately enriched with cholesterol, leading to C/PL molar ratios greater than 1.0. (From Cooper, R. A., Diloy-Puray, M., Lando, P., and Greenburg, M. S., *J. Clin. Invest.*, 51, 3182, 1972. With permission.)

lipid). The C/PL observed in membranes may reflect an average of the relative amounts of these various domains.

Membrane fluidity correlates well with the average C/PL of membranes. This has been studied using a variety of membrane probes. For example, using the fluorescent probe, 1,6-diphenyl-1,3,5-hexatriene (DPH), a close relation was observed between the C/PL of red cell membranes and membrane fluidity through a range of red cell membrane C/PL values extending from 0.5 to 2.0.[9] An increase in membrane C/PL beyond a value of 2.0 was not detectable as a further decrease in membrane fluidity. This upper limit of 2.0 in terms of membrane C/PL represents a 1:1 interaction of cholesterol molecules with the acyl chains of phospholipids. It is possible that membranes with C/PL values in excess of 2.0 have domains which include cholesterol in a form that fails to influence membrane fluidity. Alternatively, such domains may exclude DPH, and therefore the dynamics within these domains may not be represented by the rotational diffusion of this probe.

Is the symmetry of cholesterol across the bilayer altered when the amount of cholesterol is varied? Flam and Schachter have addressed this question by measuring the fluidity of the inner and outer leaflet of red cell ghosts under conditions of varying cholesterol content.[47] The inference from their studies is that when red cells are depleted of cholesterol, a disproportionate amount of that cholesterol is derived from the inner leaflet, whereas when red cells are enriched with cholesterol, a disproportionate amount of cholesterol is incorporated into the outer leaflet. Although provocative, these conclusions must be tested by independent experimental means.

The syndrome of spur cell anemia represents an abnormality of red cell membrane cholesterol content resulting from the equilibrium partition of cholesterol between abnormal lipoproteins and normal red cell membranes. It occurs in patients with severe liver disease and it results from a primary disorder of plasma lipoprotein metabolism which leads to cholesterol enrichment of lipoproteins.[48] Cholesterol partitions into red cell membranes increasing membrane C/PL by 25 to 65%, thus increasing the C/PL mole ratio to values as high as 1.6 (Figure 2).[49] It is of interest to compare this value to values which can be achieved in vitro when cholesterol is mixed with pure phospholipids. Under conditions of high lipid concentration and gentle agitation, multilamellar liposomes form in which the maximum C/PL is 1.0.[50] This value has been considered to be the upper limit of C/PL, and normal membranes have C/PL values which are below 1.0. Values > 1.0 can be achieved with sonication. [9,17,51] Thus, although a value of C/PL < 1.0 may be preferred, values of C/PL > 1.0 can be achieved in vitro and are found in disease states in vivo.

The syndrome of spur cell anemia is not unique to man, but has been described in rodents (guinea pigs and rabbits)[52,53] and in dogs[54] fed cholesterol-rich atherogenic diets. The accumulation of cholesterol is not confined to red cell membranes. In man and in dogs, platelet membranes are affected as well,[48,54] and in rodents, an increased C/PL has been observed in the surface membranes of macrophages[55] and liver cells.[56] Moreover, the C/PL of lymphoblast surface membranes increases when these cells grow as ascites tumors in cholesterol-fed mice.[57] Nor are changes isolated to surface membranes. Cholesterol equilibrates between cell surface membranes and internal membranes,[58] causing an increase in the C/PL of microsomal membranes in cholesterol-fed animals.[59] Thus, an increase in the C/PL of plasma lipoproteins leads to an increase in the C/PL of many cell membranes.

C. Equilibrium Partition of Cholesterol Plus Phospholipid

The preceding discussion is focused on the discrete movement of cholesterol, either by exchange diffusion or by equilibrium partition. The former movement does not result in a net change in sterol content, whereas the latter involves a net transfer of cholesterol. This equilibrium partition of cholesterol does not involve the simultaneous transfer of phospholipid. However, under some circumstances, a net movement of both cholesterol and phospholipid has been observed. One example is patients with liver disease. Most patients who have liver disease do not have cholesterol-rich spur cells. Rather, their red cells have an increased membrane surface area causing them to have a target-like appearance, due to the acquisition of an excess of both cholesterol and phospholipid in proportional amounts (Figure 2).[49,60] The increase in cholesterol per cell may be as great as 75%, but more commonly ranges between 25 and 50% above normal. Although there is variability from patient to patient, the percent increase in phospholipid is about two thirds of the percent increase in cholesterol, resulting in a small increase in C/PL. This increase in phospholipid content is not distributed among all phospholipids, but rather is confined to lecithin.

It is of interest that a similar abnormality in red cells has been observed in patients with a congenital absence of the plasma enzyme, LCAT.[61] LCAT deficiency of a variable degree is also quite common in patients with liver disease. However, the serum LCAT activity of patients with liver disease correlates poorly with the degree of lipid accumulation within their red cell membranes.[49]

Two processes appear to participate in determining the amount of cholesterol and lecithin in target cells. First is a process analogous to, if not identical with, the isolated transfer of cholesterol from lipoproteins to red cells as dictated by the C/PL of the lipoprotein. Since the C/PL of LDL in most patients with liver disease is increased only mildly,[49] this accounts for only a small amount of the additional membrane cholesterol. However, this process is readily demonstrable in vitro, and it accounts for the increased C/PL of target cells in liver disease.[49, 62]

The larger amount of cholesterol and lecithin which is acquired by red cells appears to involve an independent transfer of lecithin followed by a process of equilibrium partition during which the C/PL of the red cell (now transiently decreased because of this added lecithin) comes into equilibrium with the increased C/PL of LDL. This results in the transfer of cholesterol from lipoproteins to cell membranes. In this regard, it is of interest that the normal phospholipid/protein weight ratio of LDL is approximately 1.0, and this is increased 25% in patients with liver disease.[49] Possibly this elevated phospholipid/protein ratio of LDL underlies the increased lecithin content of target cells. Thus, in red cells and other cells the total content of cholesterol in membranes appears to be under at least three separate influences: (1) the membrane phospholipid content, (2) the fraction of phospholipid which is available within each membrane for solubilizing cholesterol, and (3) by the C/PL of plasma lipoproteins which are in equilibrium with cell membranes.

V. PINOCYTOSIS OF LIPOPROTEINS

The uptake of lipoprotein cholesterol by fluid phase pinocytosis, like the process of exchange diffusion, is not lipoprotein specific.[63] The amount of uptake is determined by the pinocytotic activity of the cells and the concentration of lipoprotein in the extracellular medium. At a given rate of pinocytosis, uptake varies as a linear function of the lipoprotein concentration. Unlike exchange diffusion, pinocytosis requires cellular energy. While receptor-mediated interactions are of quantitatively greater significance in terms of uptake of lipoprotein-carried cholesterol by certain cells, especially steroid hormone-producing cells, pinocytosis can make an appreciable contribution. In several species, including man, 50 to 60% of the circulating LDL is cleared by pathways not mediated by the LDL receptor.[64] In humans with homozygous familial hypercholesterolemia who lack functional LDL receptors and have an increased daily production rate of LDL,[65] the fractional rate of removal of LDL from the plasma compartment is low, but the mass flux of LDL through the plasma compartment is actually increased. Pinocytosis of LDL probably plays a significant role under these circumstances.[66] In contrast to the receptor-mediated process of lipoprotein uptake, pinocytotic accumulation of lipoproteins is not subject to feedback regulation.

VI. RECEPTOR-MEDIATED UPTAKE OF LIPOPROTEINS

In contrast to exchange diffusion and pinocytosis, receptor-mediated interactions between lipoproteins and cells are lipoprotein specific; they are also concentration dependent and saturable. Like pinocytosis, cellular energy is required for the uptake of the lipoprotein particles. Evidence for the existence of several different lipoprotein receptors has been presented. A receptor which recognizes apolipoproteins B and E, the so-called "LDL receptor",[65,67,68] and a receptor which recognizes HDL, presumably apolipoprotein A-I, are present on many different cell types.[68,69] The LDL receptor appears to be responsible for the uptake of lipoprotein cholesterol by extrahepatic cells. The functions of the HDL receptors are less well understood; in rodents they play a role in the delivery of sterol to steroidogenic cells, but in other species they may possibly serve a function in the removal of sterol. Other lipoprotein receptors have been described which are unique to specific cell types. These include a receptor specific for apolipoprotein E, which is probably restricted to liver cells,[70] and receptors for modified LDL, which are characteristic of scavenger cells such as macrophages and Kupffer cells.[71,72]

A. The LDL Pathway

Among lipoprotein receptors or putative receptors, the LDL receptor is the best characterized. It has been purified to homogeneity from bovine adrenal cortex and found to be an acidic glycoprotein of 164,000 mol wt.[67] The structure of this receptor is widely conserved across species, since antibodies raised against the bovine adrenal LDL receptor recognize similar receptors on human, rat, and canine cells.[73] The LDL receptor has a calcium-dependent high affinity for both apolipoprotein B- and apolipoprotein E-containing lipoproteins. The affinity of the receptor for lipoproteins containing apolipoprotein E is substantially higher than for lipoproteins containing apolipoprotein B. However, at saturation a single particle of E-containing apolipoprotein is bound by each LDL receptor, whereas four apo-

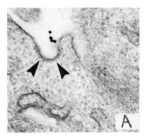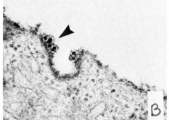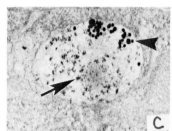

FIGURE 3.   Uptake of LDL-conjugated to colloidal gold by luteinized rat granulosa cells in vitro. A. After exposure of cells to gold-LDL conjugates at 4°C and warming to 37°C for 2 min, the conjugates (black dots) are localized over bristle-coated pits (arrow heads). (Magnification × 51,000.) B. The lipoproteins, which surround the gold particle, can be visualized after staining with ruthenium red. (Magnification × 49,500.) C. In cultures exposed to gold-LDL conjugates at 4°C then warmed to 37°C for 60 min, gold particles (arrow head) are seen primarily in lysosomes. In this case a lysosomal marker enzyme, acid phosphatase, has been co-localized to the same structure as the gold particles (arrow indicates acid phosphatase reaction product). (Magnification × 52,250.) (From Paavola, L. G. and Strauss, J. F., III, Unpublished observations.)

lipoprotein B-containing LDL particles bind to each receptor.[74,75] The arginine and lysine residues of apolipoproteins B and E seem to play an important role in the recognition process since their chemical modification prevents binding of the particles to the LDL receptor.[65,75,76]

Following the binding of lipoproteins to LDL receptors, the particles must be internalized and processed before the lipoprotein-carried sterol can be made available to the cell. These steps constitute what has been termed the "LDL pathway".[77,78] The pathway was first elucidated from studies on cultured human fibroblasts, but has subsequently been found to be a ubiquitous process both in vitro and in vivo. Ultrastructural studies have revealed that the LDL receptors cluster in specialized indented regions of the plasma membrane, referred to as "coated pits" (Figure 3). This coating is on the cytoplasmic face of the pit. It is formed by polygonal networks of the protein clathrin, which is attached to the cytoplasmic surface of the membrane in a noncovalent manner. The rearrangement of cathrin from hexagonal to pentagonal forms is thought to be the fundamental force driving endocytosis.[79] In human fibroblasts, coated pits cover approximately 2% of the cell surface, but they contain up to 70% of the LDL receptors.[78] Coated pits invaginate almost immediately after they are formed and pinch-off from the plasmalemma, thus internalizing both LDL and the receptors. The internalization process is remarkably rapid; half the surface-bound LDL of a cultured human fibroblast is internalized every 5 min. Internalization of coated pits goes on even in the absence of LDL, and the rate of receptor internalization is not affected by the presence of LDL.

Once internalized, the clathrin coating is lost from the internalized plasma membrane and LDL is delivered to lysosomes.[77,78] Lysosomal proteinases degrade LDL apolipoproteins, and lysosomal acid hydrolase releases free cholesterol from the LDL cholesteryl esters. Inhibitors of lysosomal function such as chloroquine prevent catabolism of the internalized LDL.[80] The receptors, which are separated from LDL at some point in the processing, are available for reinsertion into the plasma membrane, apparently in a random fashion. As cells do not contain a large internal pool of receptors, rapid recycling is required to maintain the cell surface receptor complement. This process of LDL receptor recycling in human fibroblasts can be interrupted by the carboxylic ionophore, monensin.[81]

B. Regulatory Functions of Cholesterol Derived via the LDL Pathway

Cholesterol derived from LDL may become integrated into membranes or it may serve as a precursor for steroid hormones or bile acids. In addition, it exerts several key regulatory effects upon cellular sterol metabolism, including suppression of *de novo* sterol synthesis and LDL receptor activity, and stimulation of cholesteryl ester synthesis.[65,77] The overall goal of these regulatory responses is to coordinate the intra- and extracellular sources of cholesterol so that cellular demands are met and the free cholesterol content of cell membranes is stabilized. In fibroblasts[77] and rat hepatocytes[82] the responses to physiological LDL concentrations described above require that the lipoprotein be accumulated via a receptor-mediated pathway; LDL cholesterol taken up via pinocytosis is not capable of exerting these regulatory effects. However, this may not be true of all cell types.[83]

1. Regulation of De Novo Cholesterol Synthesis

The *de novo* biosynthesis of cholesterol involves more than thirty discrete enzymic steps.[84] The principal, but not the only, site of regulation of sterol biosynthesis is generally acknowledged to be 3-hydroxy-3-methylglutaryl coenzyme A reductase (HMG-CoA reductase). This enzyme is subject to several types of control by a variety of regulators, most prominent of which is negative feedback by the availability of cholesterol.[85,86] A primary means of regulation of the enzyme is achieved through control of enzyme synthesis and degradation. HMG-CoA reductase has a relatively short half-life, estimated to be on the order of 2 to 3 hr.[85,86] Thus, changes in rates of enzyme synthesis readily affect enzyme concentrations. In addition to changes in enzyme content, alterations in the membrane milieu of the enzyme can affect its activity.[87] HMG-CoA reductase activity is also modulated acutely by phosphorylation/dephosphorylation, mediated through a cascade of protein kinases which inactivate the enzyme and a phosphoprotein phosphatase which reactivates it.[88] Using the phosphorylation/dephosphorylation mechanism, cells may respond rapidly with increased mevalonate synthesis. When negative feedback by cholesterol is exerted, several types of control of HMG-CoA reductase may come into play including an initial decline in activity due to phosphorylation of the enzyme followed by enzyme degradation and suppression of enzyme synthesis.[89,90]

Just how the negative feedback signal for cholesterol synthesis is detected by cells is not known. Indeed, it is uncertain whether the effector is cholesterol itself, or perhaps a metabolite of cholesterol or intermediate in the sterol biosynthetic pathway.[91] One possible means of sensing the negative feedback signal could be alterations in membrane function due to a relative increase in sterol content. However, it is evident that the feedback signal also influences rates of enzyme synthesis, indicating an action on the genome.

Recently, a great deal of attention has focused on the hydroxysterols as potential mediators of the negative feedback response. Hydroxysterols such as 7-ketocholesterol and 25-hydroxycholesterol have been shown to be particularly potent suppressors of HMG-CoA reductase in cultured cells.[91] These hydroxysterols are more soluble in water than cholesterol,[92] and are readily taken up by cells. They can mimic all the actions of LDL-carried cholesterol, including suppression of LDL receptor activity and stimulation of cholesteryl ester synthesis.[93,94] Hydroxysterols occur in mammals, including man, as metabolites or oxidation products of cholesterol,[95] and it has been postulated that cells might contain specific binding proteins for such regulating sterols. By analogy to the mechanism of steroid hormone action, these binding proteins might serve as "receptors", allowing the hydroxysterol-receptor complex to exert effects on the genome and regulate expression of the HMG-CoA reductase. While cytosolic binding proteins having a relatively high affinity for 25-hydroxycholesterol have been described by several investigators,[96,97] it remains to be seen if they have any direct function in the regulation of HMG-CoA reductase, or any other enzyme in the sterol biosynthetic pathway.

It should be noted that HMG-CoA reductase is subject to multivalent feedback control, not simply regulation by availability of cholesterol or a cholesterol metabolite.[86] In addition to sterols, at least three other molecules essential for cell function and growth are derived from mevalonate: ubiquinone, dolichol, and isopentenyl adenine. Liopoprotein-derived cholesterol is not capable of completely suppressing HMG-CoA reductase activity, and a nonsterol regulator, which is normally synthesized endogenously from mevalonate, is required to give full enzyme suppression in the presence of exogenous sterol.[86] Thus, the nonsterol regulator appears to act cumulatively with sterol to regulate HMG-CoA reductase levels. This multivalent feedback process in part prevents sterols from blocking the synthesis of the other essential isoprenoid molecules.

While HMG-CoA reductase is believed to be the rate-limiting enzyme in cholesterol synthesis, other enzymes in the pathway are affected by cholesterol availability in a negative feedback fashion. These include HMG-CoA synthetase,[98] a cytosolic enzyme which catalyzes the reaction immediately prior to HMG-CoA reductase, and several later steps such as squalene synthetase,[99] sterol demethylation and reduction of the Δ^{24} double bond.[100] Alterations in these enzymes may modulate flux rates, and in the case of squalene synthetase, direct precursors toward the synthesis of the essential nonsterol isoprenoid products in times of adequate supply of exogenous cholesterol.

2. Regulation of LDL Receptors

The negative feedback of cholesterol on cellular sterol metabolism extends to the LDL receptor.[93] Both cholesterol delivered by the LDL pathway and hydroxysterols reduce the number of LDL receptors by suppressing receptor synthesis. Unlike HMG-CoA reductase, LDL receptors have a relatively long half-life on the cell surface ($\sim$ 25 hours). It appears likely that the same signal which results in the suppression of HMG-CoA reductase synthesis causes a reduced rate of LDL receptor formation.

3. Regulation of Cholesteryl Ester Storage

Cholesterol derived from the LDL pathway or cholesterol in excess of cellular requirements obtained by other means stimulates cholesteryl ester storage. This diversion of cholesterol into cholesteryl esters may protect the cell membranes from an excess of cholesterol which it might otherwise have to endure.[101] The amount of cholesteryl ester stored in cells is affected not only by synthesis, but also by hydrolysis. This pool of cholesterol, or a portion of it, seems to turn over constantly, and the rate at which free cholesterol enters into the pool in relation to the rate of sterol ester hydrolysis determines whether the pool expands or shrinks.[102]

Cholesteryl ester synthesis is catalyzed by an endoplasmic reticulum enzyme, acyl CoA: cholesterol acyltransferase (ACAT). This enzyme esterifies long chain fatty acyl CoA and cholesterol, and the cholesteryl esters so formed are stored as lipid inclusions.[103] The rate of sterol esterification seems to be governed by the entry of free cholesterol into the substrate pool.[101,104,105] This process may be affected by transport proteins which facilitate movement of sterol from membranes to the appropriate compartment in the endoplasmic reticulum. Sterol carrier protein$_2$, also called nonspecific phospholipid transfer protein, has been shown to increase esterification of free cholesterol by ACAT under in vitro conditions.[106]

The primary factors regulating mobilization of sterol ester from cellular lipid droplets appear to be the activity of a cytoplasmic enzyme, sterol ester hydrolase, and the availability of an appropriate acceptor or metabolic process to remove cholesterol from the cell.[102] In certain cell types, most notably steroidogenic cells, the activity of the sterol ester hydrolase is increased acutely as a result of phosphorylation catalyzed by a cyclic AMP-dependent protein kinase following exposure of the cells to tropic stimuli.[107] Such stimuli also enhance the metabolism of cholesterol into steroidal hormones.

It is noteworthy that the process of cholesteryl ester storage is costly in terms of cellular energy.[102] First, cholesterol is primarily taken up in the esterfied form. It is hydrolyzed by lysosomal acid lipase and must then be re-esterified at the expense of an activated fatty acid for storage. Second, the stored sterol esters constantly turn over, creating a futile cycle of synthesis and hydrolysis which consumes ATP through the requirement of activated fatty acid for sterol ester synthesis.

4. Cellular Cholesterol Balance: Normal and Disturbed

Fibroblasts and lymphocytes represent simple models for study of the regulation of cellular cholesterol balance.[65] In the presence of an adequate supply of LDL, they establish a steady state in which *de novo* sterol synthesis is suppressed, and the small amount of cholesterol the cells need is acquired either via the few LDL receptors or by means of exchange diffusion. Excessive cholesterol expands the cellular sterol ester pool. If the cells are deprived of LDL or increase their metabolic needs for cholesterol, such as during cell division, the sterol ester pool is depleted and both HMG-CoA reductase and the number of LDL receptors increase, creating a new steady state. In steroidogenic cells, tropic hormones also appear to influence the expression of lipoprotein receptors, thus assisting these cells in meeting their greater need for cholesterol.[108]

The regulation of cholesterol metabolism in human cells can be disturbed by a variety of genetic disorders.[65,109] These include defects in the expression or function of LDL receptors, defects in the internalization of lipoprotein receptors, and deficiency of enzymes involved in lipoprotein processing.[110] Mutations affecting LDL receptor function, the most common of which is absence of the receptor, produce familial hypercholesterolemia in man.[65] These types of mutations also occur in animals. The Watanabe heritable hyperlipidemic rabbit appears to have a defect identical to human familial hypercholesterolemia.[111] All of these disorders render cells incapable of taking up and processing cholesterol via the LDL pathway and, therefore, require the cells to generate sterol for metabolic needs by *de novo* synthesis or by uptake via nonreceptor-mediated mechanisms. The regulatory system in these cells which perceives the negative feedback signal of cholesterol is, however, intact and can be brought into play by providing the cells with 25-hydroxycholesterol.

In certain tumor cells the expected negative feedback control of cholesterol over *de novo* sterol synthesis is lost.[112,113] This phenomenon has been demonstrated in hepatomas as well as leukemic cells. Indeed, in every hepatoma thus far investigated, the feedback relationships are absent or defective.[112] The loss of feedback regulation in hepatomas can often be demonstrated prior to any physical manifestation of the presence of the tumor. This loss of regulation may lead to an accumulation of cholesterol in cellular membranes of affected cells and, subsequently, altered membrane function which may promote tumorogenesis. However, this may also be related to the enhanced rate of cellular proliferation since *de novo* sterol synthesis is a distinct metabolic requirement for certain cells before they enter the proliferative cycle.[114] As the nature of the control in normal cells is not yet known, the alterations which lead to aberrant function in precancerous and cancerous tissues also remain to be elucidated.

Using cell culture techniques,[115] it has recently been possible to isolate a number of mutants which are defective in sterol synthesis in both the early (HMG-CoA synthase)[116] and late (sterol demethylase)[117] stages of the biosynthetic pathway. Of greater interest, however, are mutants which display defects in the regulation of sterol synthesis. Sinensky and his colleagues[115,118] have described cell lines derived from Chinese hamster ovary cells which are resistant to the cytotoxic action of 25-hydroxycholesterol when grown in lipoprotein-depleted medium. Some of these cell lines display defects in the negative feedback response to hydroxysterol and are constitutive for the expression of HMG-CoA reductase. Other mutants appear to have defects in HMG-CoA reductase degradation. These cell lines will afford investigators a means of probing the molecular nature of the feedback control. One significant observation which has already come from the study of these mutant cells is that a defect in the control of *de novo* sterol synthesis can occur without a change in rates of cellular uptake of cholesterol, cholesterol esterification, or rates of cholesterol efflux. These findings suggest that each of the processes regulating cellular free cholesterol levels can be independently regulated, even though under usual circumstance they undergo coordinate changes.

C. The Interaction of HDL with Cells

Evidence exists for a distinct binding site on the surface of cells which recognizes HDL by a process other than the recognition of apolipoprotein E. HDL deficient in apolipoprotein E binds to fibroblasts, hepatocytes, and a variety of human and rat steroidogenic cells in a saturable fashion.[68,69] In contrast to the LDL receptor, the binding site for HDL is relatively insensitive to proteolytic enzymes, binding does not require calcium, and polyanions do not inhibit HDL binding.[68,119] These binding sites display a relative specificity for HDL in that unlabeled LDL is less effective than unlabeled HDL in competing for binding with labeled HDL.[68,119]

The specific determinants on the HDL particle recognized by this receptor are being studied. There is some evidence that apolipoprotein A-I, the major apolipoprotein of HDL, is the component which is recognized. Purified apolipoprotein A-I reconstituted into phospholipid vesicles binds to membranes in a fashion similar to native HDL.[119] In contrast to LDL and its receptor, modification of HDL apolipoproteins by acetylation does not affect their recognition, and the HDL binding sites do not seem to cluster in coated pits.[68,120]

In the case of rodents, the HDL receptors appear to function in the uptake of cholesterol into steroidogenic glands.[68] Moreover, tropic hormones which stimulate steroidogenesis in the adrenals, ovaries, and testes increase HDL uptake by these glands. However, the mechanism by which the HDL cholesterol is taken into such cells remains unclear. Relatively little degradation of the HDL apolipoproteins occurs despite the utilization of the HDL-derived cholesterol by the cells.[68,121] Possibly, HDL is taken up in noncoated vesicles after which cholesterol is liberated and apolipoproteins are regurgitated from the cell.[122] Alternatively, cell surface-associated lipolytic activity identical to "hepatic lipase" may hydrolyze HDL phospholipids, thereby increasing the C/PL mole ratio of the particle and favoring a net movement of cholesterol into the cells.[123] In this regard, the rat adrenals and ovaries are known to contain substantial hepatic lipase activity.

In species other than rodents, the function of the HDL binding sites is not well understood. In contrast to the LDL receptor, mutations affecting HDL binding sites have not yet been discovered, and there are no known disorders related to HDL binding. In the liver, these receptor sites might play a role in uptake of HDL, but direct evidence for this is not available. In nonhepatic cells, it is likely that these receptors do not participate in the uptake of HDL cholesterol, as HDL is not usually capable of meeting cellular demands for exogenous sterol.[69] In fact, exposure to HDL often stimulates *de novo* sterol synthesis, suggesting that these particles facilitate sterol efflux (vide infra). The metabolism of HDL by extrahepatic tissues in species other than rodents is also quite different than the metabolism of LDL. While fibroblasts as well as other cell types have a number of HDL binding sites equivalent to the number of LDL receptors, the rate of internalization and degradation of the HDL apolipoproteins is markedly less than that for LDL.[124-126] These characteristics are consistent with a role for HDL in the efflux of sterol from cells.

Little is known about the regulation of HDL binding sites in species other than rodents, except that their control is distinct from that of the LDL receptor. In fibroblasts from patients with familial hypercholesterolemia which are deficient in LDL receptors, HDL binding sites are slightly increased.[125] In normal cells, incubation with hydroxysterol reduces LDL receptors without affecting HDL binding sites.[126]

VII. EFFLUX OF STEROLS FROM CELLS

Besides delivering cholesterol to cells, lipoproteins play an important role in removing cellular cholesterol. Using a variety of different cell types in culture, it has been found that lipoproteins, particularly HDL, can promote significant excretion of free sterol.[69,127-129] Esterified sterol can also be removed from cells, but prior hydrolysis of the ester bond is required before the sterol can be released. The capacity to stimulate sterol efflux is expressed by native HDL particles as well as by HDL apolipoproteins reconstituted into phospholipid vesicles. The C/PL mole ratio in HDL particles, particularly HDL_3, is believed to modulate the process of sterol efflux. The events in cholesterol removal appear to be the reverse of those described for exchange diffusion of cholesterol from lipoproteins to cells.[127] Studies in model systems suggest that free cholesterol molecules leave the plasma membrane and partition into the aqueous phase where they can subsequently collide and be absorbed by an acceptor. The overall rate of efflux seems to be determined by the desorption of cholesterol from the plasmalemma when there is a sufficient concentration of acceptor in the cellular environment. While data obtained from the study of model systems in vitro can be interpreted without invoking interactions of cholesterol acceptors with cell surface receptors, in vivo the association of HDL with specific binding sites on the cell surface might facilitate the removal of sterol from cells by increasing the concentration of acceptor particles in the cell environment.

Factors in addition to HDL are required for efficient cholesterol efflux in vivo. These include LCAT, a plasma enzyme which is activated by apolipoprotein A-I of HDL and which catalyzes the esterification of HDL free cholesterol utilizing fatty acyl moieties from phosphatidylcholine, and plasma cholesteryl ester exchange/transfer proteins which facilitate the movement of HDL cholesteryl esters to other lipoproteins, particularly LDL.[130] By these mechanisms, a C/PL mole ratio favorable for free cholesterol uptake by the HDL particle can be maintained.

While the concept that HDL acts as the primary acceptor of cellular cholesterol is appealing, it may be quite simplistic. Recent studies on virus-transformed human lymphoblastoid cells point out some of the complexities of HDL-cell interaction.[131] It was found that HDL_3 at concentrations below 100 μg/mℓ stimulated cellular *de novo* sterol synthesis, diminished cholesteryl ester formation, and increased cellular cholesterol content under conditions where sterol efflux was not significant. At higher concentrations of HDL, *de novo* sterol synthesis was not stimulated but the cells appeared to take up cholesterol from HDL. These dose-related actions of HDL are at variance with older observations and thinking; they reveal that our understanding of HDL-cell interactions is far from complete.

Cellular cholesterol excretion may also be facilitated by the actual synthesis and secretion of apolipoproteins by peripheral cells. Recent studies reveal that mouse peritoneal macrophages[132] as well as human kidney and adrenal glands[133] synthesize apolipoprotein E or a protein remarkably similar to it in appreciable quantities (0.2 to 2% of total protein synthesis). In the case of the mouse macrophage, synthesis was related to the cellular sterol content. These provocative observations raise the possibility that peripheral production of apolipoprotein E may play an important role in lipid transport and metabolism. The secretion of apolipoprotein E by peripheral cells might serve to direct excreted cholesterol to the liver where it could be cleared by the apolipoprotein E-specific receptor.

VIII. CHOLESTEROL EFFECTS ON MEMBRANE FUNCTION

The complex systems through which cell cholesterol is regulated strive to preserve a constant membrane C/PL mole ratio. Changes in the C/PL mole ratio of cell membranes have a number of important effects on cells. These include changes in permeability and transport functions, changes in membrane enzyme activity and in the availability of membrane components as substrates, changes in protein conformation and receptor position, and changes in both cell shape and in the ability of cell membranes to undergo shape changes. Examples of each will be presented briefly.

A. Permeability and Transport

An effect of cholesterol on permeability was first demonstrated using lecithin liposomes containing various sterols, and these studies helped to establish the structural requirements for sterols in membranes.[134] Although not all studies utilizing red cells are in agreement, it appears that increases in membrane cholesterol reduce the passive permeability and facilitated diffusion of a number of electrolytes and nonelectrolytes, whereas cholesterol depletion tends to increase permeability. Thus, cholesterol depletion increases the passive permeability of red cells to glycerol, acetate, and Na^+.[17,135] An apparent increase in active Na^+ flux also accompanies cholesterol depletion. Although this may be explained in terms of a response to the increase in passive Na^+ permeability rather than a primary effect on the transport mechanism,[17] an effect of cholesterol on the Na^+ affinity of the Na^+-K^+ pump has also been demonstrated.[136] In contrast, cholesterol enrichment inhibits the furosemide-sensitive diffusion of $Na^+ + K^+$,[137] as well as the diffusion of glycerol, erythritol, acetate, and proprionate.[138] Similar decreases in nonelectrolyte permeability have been observed in cholesterol-rich guinea pig red cells.[61] We have not observed any effect of cholesterol enrichment on active Na^+ or K^+ transport,[17] although a decrease in active Na^+ flux of small magnitude has been reported in cholesterol-rich guinea pig red cells.[139] Thus, modulation of membrane cholesterol has a substantial effect on the permeability of the cell to small molecules.

B. Membrane Enzymes

The effect of cholesterol on membrane enzymes has been analyzed in rats fed a corn oil diet supplemented with cholesterol.[140] Cholesterol enrichment affected acetylcholinesterase and Na^+-K^+ ATPase, while causing no change in Mg^{2+}-ATPase. The inclusion of cholesterol in reconstituted vesicles containing Ca^{2+}-ATPase from sarcoplasmic reticulum caused a decrease in this enzyme activity.[141] These effects on membrane enzymes may relate to the lipid composition of the immediate environment of the enzyme. Studies with both cytochrome oxidase from mitochondria[142] and Ca^{2+}-ATPase from sarcoplasmic reticulum[143] have demonstrated a specific boundary layer, or annulus, of phospholipid. Moreover, it appears that cholesterol is specifically excluded from the annulus in sarcoplasmic reticulum. Therefore, in a manner quite distinct from its bulk influence on membrane fluidity, cholesterol may influence membrane enzymes by directly interacting with these boundary lipids.

1. Receptors

Dr. Shinitzky has presented evidence that enrichment of red cell membranes with cholesterol causes an increase in the exposure of certain membrane proteins to the aqueous environment.[144,145] One manifestation of this phenomenon is an enhancement of the exposure of the Rh D antigen in red cells rendered cholesterol-rich by prior incubation with liposomes containing high C/PL molar ratios.[146] In contrast, neither the mobility of band 3 protein nor its susceptibility to protease digestion was affected when red cell membranes were enriched with cholesterol.[147,148] However, following proteolytic digestion of the bonds which associate band 3 molecules with cytoskeletal components, a direct effect of membrane C/PL mole ratio on band 3 mobility was demonstrated.[149] Thus, the position and mobility of some membrane proteins may be sensitive to the C/PL of their environment.

C. Membrane Components as Substrates

Studies of human platelets have provided insight into the possible role of membrane C/PL mole ratio in modulating the availability of membrane components as substrates. When

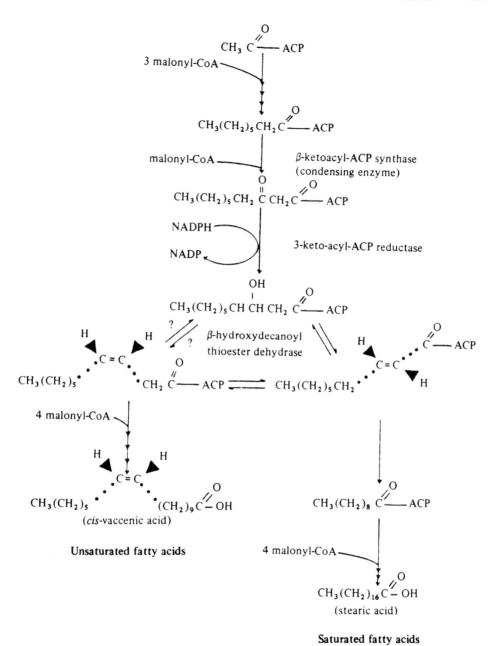

FIGURE 2.   Fatty acid desaturation in *E. coli*. (Modified from Thompson, G. A., Jr., *Regulation of Membrane Lipid Metabolism*, CRC Press, Boca Raton, Fla., 1980.)

The activity of the β-hydroxydecanoyl thioester dehydrase is thought to be a primary factor in regulating the unsaturated to saturated fatty acid ratio under these conditions, although recent studies by Cronan and his colleagues[24,25] have indicated that this enzyme is only one of several functions which serves to regulate fatty acid composition with respect to temperature. These workers reported evidence for the existence of a separate enzyme, called the β-ketoacyl synthase II, which is responsible for the conversion of palmitoleic acid to *cis*-vaccenic acid.[24,25] This enzyme is controlled by a separate genetic locus from the synthase I which is responsible for elongation of the acyl chains in the pathway through the

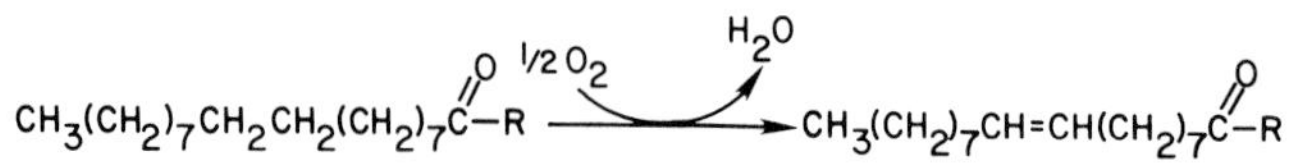

$$CH_3(CH_2)_7CH_2CH_2(CH_2)_7C\overset{O}{-}R \xrightarrow[\hspace{1.2cm}]{\overset{\textstyle 1/2\,O_2\quad H_2O}{\nearrow\searrow}} CH_3(CH_2)_7CH=CH(CH_2)_7C\overset{O}{-}R$$

FIGURE 3. Generalized fatty acid desaturation in aerobic cells. (Modified from Thompson, G. A., Jr., *Regulation of Membrane Lipid Metabolism*, CRC Press, Boca Raton, Fla., 1980.)

16-carbon species. The activity of the synthase II gene seems to be responsible for much of the temperature-dependent increase in unsaturated fatty acids. Mutants have been isolated at the fabF locus, which appears to be the structural gene for the enzyme.[26] These are unable to convert palmitoleic acid to *cis*-vaccinate and in addition do not increase the content of total unsaturated acids following a shift from high to low temperatures. Recently the Cronan group has reported on another mutant designated Vtr which maps near the fabF locus. This latter mutation also appears to affect synthase II activity and results in the overproduction of vaccenic acid.[27]

In efforts to shed light on the enzymatic basis behind the temperature regulation of fatty acid biosynthesis, Okuyama and colleagues[28] have reported on in vitro studies of fatty acid synthetase systems isolated from *E. coli* cultures grown at 40°C and 10°C. When both systems were assayed at either 40° or 10° the synthetase preparation from the high temperature cultures was found to produce fatty acid mixtures with a higher ratio of unsaturates to saturates and with longer chain lengths than made by the synthetase from the low temperature cultures.[28] Synthetases from the two culture conditions also showed differing responses to malonyl-CoA concentrations, indicating that adaptation to lower temperatures involves some physical change in the fatty acid synthetase complex. It is not clear at this point which of the enzyme components within the system have been modified.

2. Desaturation in Aerobic Organisms

In contrast to the anaerobic situation where unsaturated bonds can be incorporated during the initial biosynthetic process, most organisms introduce double bonds into preformed long chain saturated acyl-CoA species by an oxygen-dependent, iron-requiring process. The basic features of oxygen-mediated desaturation are summarized in Figure 3. All systems appear to have an absolute requirement for oxygen, and the natural electron donors are reduced pyridine nucleotides. In most systems, electrons appear to be transferred from pyridine nucleotides to the active site of the desaturase through a membrane-bound electron transport chain.

Studies of oxygen-dependent desaturation mechanisms have been hampered by difficulties in solubilizing and fractionating these enzyme systems into their individual components. Furthermore, even the identity and pool sizes of desaturase substrates and products have been troublesome to establish. There is general agreement that the initial substrate is a saturated acyl-CoA, but the competition of other enzymes, e.g., fatty acyltransferases and fatty acyl-CoA hydrolases for saturated acyl-CoAs and their desaturation products, mono-unsaturated acyl-CoAs, makes estimation of their normal rates of in vivo metabolism difficult.[19]

Almost all in vitro systems studied to date appear to rapidly desaturate acyl-CoA derivatives. It was originally thought that fatty acyl-CoAs were in fact the true substrates for desaturation. This may in fact be the case for most bacteria, but recent studies with higher organisms have brought forth convincing evidence that the acyl chains of intact phospholipid molecules can be desaturated *in situ*.[29] In plants there also appears to be a desaturation system specific for ACP derivatives.[30] Free fatty acids do not appear to be utilized by desaturases.

Given the diversity of desaturase substrates, both structurally and in their physical properties, determining the exact nature of the substrate in a particular case is of considerable importance in understanding how that reaction is regulated. Desaturation of acyl-CoA and acyl-ACP derivatives, for example, could be envisioned as occurring at or near the membrane surface in a relatively hydrophilic environment, whereas direct desaturation of phospholipid fatty acids would most likely occur within the hydrophobic region of the bilayer.

a. Desaturation in Aerobic Bacteria

In aerobic bacteria, generally only one double bond is inserted into fatty acids. Palmitoleic and oleic acids are common products of desaturation, although other species are also known to occur in nature.[20] *B. megaterium*, for example, converts palmitate and stearate to *cis*-5 derivatives whereas *M. phlei* desaturates palmitate primarily at the 10 position.

The most extensively studied desaturation system in aerobic bacteria is that of *B. megaterium*. Work by Fulco and his co-workers[31] has shown that the Δ5 desaturase activity in this organism is under sophisticated control which results in the production of high levels of unsaturates when the bacterium is grown at low temperatures. Regulation of desaturase activity within the cells is governed by several temperature-dependent factors. In isothermally growing logarithmic phase cells, steady-state levels of the enzyme appear to be maintained primarily by turnover of the desaturase system caused by factors which irreversibly inactivate the enzyme. Thus, at intermediate temperatures, a 2° decrease in growth temperature results in a doubling of the half-life of the enzyme,[20] producing higher intracellular levels of the desaturase activity.

These cells respond to shifts from high to low temperatures with a rapid increase in desaturase synthesis, which results in unusually high levels of desaturase activity. This is followed by an eventual decline to levels normally seen at the lower temperature. This "hyperinduction" phenomenon is blocked by both RNA and protein synthesis inhibitors and thus appears to differ from *E. coli* in requiring protein synthesis to increase enzyme activity as opposed to direct effects of temperature on preexisting enzymes.

Experiments in which exogenous monounsaturated fatty acids were incorporated into phospholipids prior to the temperature shift seem to rule out the possibility that the decay in hyperinduced activity is caused by inhibition by the newly synthesized fatty acids or by the changes they might produce in fluidity which would act on the desaturase enzyme.[31] Current evidence suggests that maintenance of steady-state desaturase levels in the cell is effected by a temperature-sensitive modulator protein which is thought to act by inhibiting synthesis of desaturase mRNA.[31] The hyperinduction of desaturase synthesis in freshly chilled cells is seemingly due to a delayed formation of functional modulator protein complexes.

b. Desaturation in Aerobic Eukaryotes

The molecular mechanisms for desaturation in eukaryotic cells are similar to those found in aerobic bacteria in that the reactions are oxygen-mediated and involve the participation of a specific electron transport chain. Unlike the bacteria, however, the predominant fatty acids in eukaryotic cells are polyunsaturated species. Needless to say, the introduction of multiple double bonds with varying positional specificities adds to the complexity of the system.

Eukaryotic cells contain a number of distinct desaturase enzymes, each of which has its own positional specificity. The most common of these, the Δ9 desaturase, (EC 1.14.99.5), inserts a double bond between carbons 9 and 10 (from the carboxyl end) of saturated fatty acyl-CoA derivatives. The fact that this specificity is maintained with substrates ranging from 10 to 22 carbon atoms in length suggests that the enzyme determines the placement of the double bond by binding the fatty acid at its carboxyl group. In spite of this specificity, the enzyme does prefer certain chain length substrates, with the C_{18} precursor being the most reactive in most plant and animal systems.

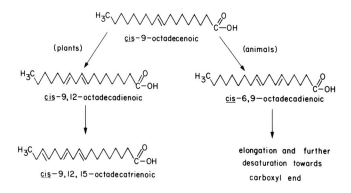

FIGURE 4.   The placement of double bonds during fatty acid desaturation in plants and animals.

Although there are exceptions, especially in the lower phyla,[20] as a general rule the insertion of additional double bonds in algae and higher plant cells is made between the first double bond and the terminal methyl group, generating a series of double bonds interrupted by methylene groups. Animal cells, however, generally tend to insert the second and subsequent double bonds between the original double bond and the carboxyl group. Thus the double bonds in plant polyunsaturated fatty acids bound to phospholipids are grouped toward the tail of the molecule whereas those in animal cells are grouped near the glycerol backbone of the phospholipid (Figure 4). The fatty acid composition in animal cells is also profoundly influenced by the availability of linoleic and linolenic acids derived from dietary plant material. These plant acids are often elongated and further desaturated by the animal prior to their incorporation into structural lipids. The fact that double bonds distort the molecule at different depths in the bilayer undoubtedly has a significant if subtle effect on the physical properties of a given phospholipid molecular species.

i. Desaturation in Animal Cells

The most detailed information concerning the components of animal desaturase systems comes from studies performed on rat liver microsomes. In these membranes the active sites of the components appear to be oriented towards the cytoplasmic side of the endoplasmic reticulum.[32] Electron flow is mediated by a flavoprotein, cytochrome b_5, and a cyanide-sensitive factor[33] which is the desaturase proper. Purified cytochrome b_5 is an amphipathic protein with a molecular weight of 16,700[34] which is attached to the membrane by a hydrophobic tail approximately 44 amino acids long, whereas the stearoyl-CoA desaturase is an extremely hydrophobic molecule consisting of a single polypeptide containing one atom of nonheme iron.[35]

Microsomal systems from rat liver have been shown to effectively desaturate both fatty acyl-CoA derivatives and acyl chains of intact phospholipids. Using synthetically prepared ^{14}C-labeled phospholipids as substrates, Pugh and Kates[36] were initially able to demonstrate that ^{14}C-labeled eicosatrienoic acid (20:3) in the *sn*-2 position of phosphatidylcholine was desaturated by rat liver microsomes to arachidonic acid (20:4). Further studies in which microsomal desaturases were solubilized by detergents, purified approximately sevenfold, and reconstituted in the presence of deoxycholate along with components of the cytochrome b_5 transport chain provided additional evidence for the direct desaturation of the phospholipid-bound substrate without conversion to the acyl-CoA ester.[29]

In eukaryotes there are numerous indications that regulation of desaturation occurs through changes in desaturase activity. A number of studies have shown that mammalian tissues such as rat liver exhibit regulation of desaturase activity under conditions where the animals are fed massive amounts of unsaturated fatty acids.[5] The existence of a large pool of unsaturated fatty acids in storage lipids of normal tissues, however, brings into question the significance of these mechanisms in regulating membrane fluidity.[5]

Fluidity-induced regulation of desaturase activity has been observed, however, in the protozoan *Tetrahymena pyriformis*. In logarithmic growth phase, fatty acids synthesized by *Tetrahymena* are almost exclusively incorporated into membrane phospholipids; therefore, events regulating the biosynthetic and modification pathways are exclusively concerned with the production and maintenance of phospholipids. Detailed studies on the control of fatty acid desaturation in this organism have shown that phospholipids from functionally distinct membranes contain characteristic fatty acid profiles which are precisely regulated over a broad temperature range.[37-39] Rapid chilling of the cells strongly inhibits *de novo* fatty acid synthesis, but fatty acid desaturation continues, leading to an elevated level of unsaturation in phospholipid fatty acids.[40] Unlike some plant systems (see Section III.B.2.b.ii), the active *Tetrahymena* desaturase activity at low temperature is not caused simply by the increased solubility of oxygen in the chilled medium.[41] Using freeze-fracture electron microscopy,[40] electron spin resonance,[38] and fluorescence polarization[42] methods, it was found that the relatively high desaturase activity was closely correlated with an increase in intrinsic viscosity of the microsomal membranes caused by the sudden chilling. Addition to such cells of fluidizing compounds such as polyunsaturated fatty acids,[40] fatty acid analogues with methoxy side chains,[43] or general anesthetics[44] inhibited desaturase activity, whereas rigidifying factors had the opposite effect. Experimental observations in this case are consistent with the idea that preexisting desaturase molecules are activated when the fluidity of their membrane environment drops below a threshold level. Consequently, the desaturase system appears to provide a self-regulating means of maintaining optimal membrane fluid properties in the endoplasmic reticulum.

An alternative explanation for the above findings would be the low temperature-induced synthesis of additional fatty acid desaturase molecules. Cells in which protein synthesis was freshly inhibited with cycloheximide still responded to chilling by substantially increasing their fatty acid unsaturation.[45] However, detailed study of the system by Nozawa and coworkers[46,47] has shown that chilling does indeed induce the synthesis of two desaturases, namely, those acting on palmitoyl-CoA and stearoyl-CoA. There is evidence that the induction features increases in components of the microsomal electron transport chain as well as the desaturases themselves. Despite a thorough search, no evidence has yet been found for the induced synthesis of the other participating desaturases. Taken together, the available evidence from *Tetrahymena* favors a dual system in which both the number of certain desaturases and the level of activity of others are regulated.

ii. Desaturation in Higher Plants

Early studies suggested that plants may have two types of desaturation systems with distinctive intracellular locations and substrate requirements. This was first noticed in studies of the green alga *Euglena*[48] which indicated that phospholipids from dark-grown, chloroplast-free cultures contained typical animal fatty acids such as arachidonic acid, while cells grown in the light (and therefore containing chloroplasts) synthesized plant-like fatty acids such as linoleic and α-linolenic acid. The latter fatty acids were primarily esterified to galactosyl glycerides, lipids normally found only in chloroplast membranes. These studies showed that while dark-grown *Euglena* could only desaturate stearoyl-CoA, light-grown cells required stearoyl-ACP as a substrate for desaturation.[49] The latter, stearoyl-ACP-requiring pathway has now been identified in the chloroplasts and proplastids of a large number of species and appears to be the major, if not the exclusive source of oleic acid in higher plants.[50,51] These findings have been recently reviewed.[52,30]

In addition to their separate intracellular locations, there are other distinct differences in the physical characteristics of plant and animal enzymes for stearate desaturation. In plants, the stearoyl-ACP desaturase systems in all tissues examined thus far consist of a soluble protein located within the proplastid or the stroma of the chloroplasts.[53] Rat liver stearoyl-CoA, by contrast, is an extremely hydrophobic membrane-bound protein located in the endoplasmic reticulum (see Section III.B.2.b.i). Enzymatic activity of the plant desaturase is specific for stearoyl-ACP. Specificity is determined by the fatty acid chain length, since ACP from *E. coli* will substitute for plant ACP in most systems. Plant desaturase systems also differ from animal and yeast systems in requiring ferridoxin rather than cytochrome b_5 as an intermediate electron carrier.[50]

Characteristics of the plant enzymes and their substrates involved in the subsequent desaturation of oleic acid to linoleate and α-linolenate are less clear. These enzymes are membrane-bound but not associated with the chloroplast, leading most observers to conclude that following the initial desaturation of stearoyl- to oleoyl-ACP, the 18:1 fatty acid is transported from the chloroplast to the endoplasmic reticulum for further modification.[30]

There is some controversy over the nature of the substrates involved in polyunsaturated fatty acid synthesis by higher plants. In the reaction involving conversion from the oleoyl to linoleoyl species, a number of workers have reported desaturation of oleoyl-CoA in in-vitro microsomal systems of safflower seeds and leaf microsomes.[54,55] Desaturation was found to be highly specific for oleoyl-CoA derivatives and required NADH and molecular oxygen. It was observed in these studies that both the oleic acid substrate and the linoleic acid product were rapidly incorporated into the 2 position of endogenous phosphatidylcholine, raising the question as to whether the oleic acid was desaturated as the CoA ester or following incorporation into phospholipid. Stymne and Appelqvist[56] provided supporting evidence for the latter possibility by showing that a safflower microsomal system effectively desaturates 2-oleoyl-phosphatidylcholine.

The nature of the substrate involved in the desaturation of linoleic acid is also in some doubt. There is agreement among a number of investigators that the preferred substrate for desaturation is a complex phospho- or glycolipid. These conclusions are based primarily on results of in vivo pulse chase studies in which exogenously supplied radiolabeled fatty acids and fatty acid precursors have been added to tissues to observe the time course of incorporation and desaturation of fatty acids into phospho- and glycolipids. In most of those studies, ^{14}C oleic and linoleic acid were found to be rapidly incorporated into phosphatidylcholine followed by subsequent conversion of the label to linolenic acid. The problem with determining the exact substrate acted on by the desaturase is that, during the same time period, phosphatidylcholine (PC) also appeared to be a precursor for the formation of the glycolipids monogalactosyl diglyceride (MGDG) and digalactosyl diglyceride (DGDG). Although a number of earlier studies have suggested that the preferred substrate for desaturation is 2-linoleoyl-phosphatidylcholine, more recent studies have indicated that the substrate may in fact be a glycolipid. Ohnishi and Yamada,[57] by following the course of incorporation of ^{14}C acetate, ^{14}C oleic acid, and ^{3}H glycerol into *Avena* leaves during greening of etiolated seedlings, have concluded that the major pathway of α-linolenate formation is 18:1 → 18:1 PC → 18:2 PC → 18:2 MGDG → 18:3 MGDG. Other evidence suggesting that phosphatidylcholine is not the actual substrate for the final desaturation step comes from the inability to detect 2-linolenoyl phosphatidylcholine under in vivo conditions.[58,59]

As with animals, plants are also capable of regulating their phospholipid fatty acid patterns in response to environmental changes as well as during development. The response of plants to temperature stress by increasing unsaturated fatty acid levels in membrane lipids following to exposure to low temperatures is thought to be a part of the frost hardening phenomenon.[60] Harris and James[61] have shown in tissues such as seeds and bulbs, where dissolved oxygen supplies are low, that the concentration of dissolved oxygen can regulate the rate of desaturation. This control was proposed to operate simply because oxygen, a cosubstrate in the desaturation reaction, is much more soluble in tissue water at low temperature and therefore stimulates the reaction by mass action. A similar dependence of nonphotosynthetic sycamore cells grown in culture on the concentration of dissolved oxygen for fatty acid desaturation has been reported by Rebeille et al.[62] Oxygen levels in photosynthetic tissue are apparently too high, due to endogenous production of the gas, to be rate limiting in the desaturation process.

Unsaturated fatty acid production is also stimulated by light in some plant tissues. Illumination of etiolated cucumber seedlings following germination in the dark stimulates desaturation of oleic acid to polyunsaturates, especially α-linolenic acid.[63] The rapidly increased production of this species is thought to be associated with chloroplast formation, since this fatty acid can comprise up to 90% of the total fatty acid in the organelle. The increased ability to desaturate labeled oleic and linoleic acids persists even when preilluminated seedlings are assayed in the dark, indicating that the stimulation is not due a direct light-induced uptake of substrate or an increase in oxygen levels produced through photosynthesis. Furthermore, increased desaturation is blocked by cycloheximide, suggesting the induced synthesis of new enzyme or modulating proteins. Light induction of desaturase activity may be limited to only certain species, however, since studies on maize seedlings did not show similar increases in activity.[59]

IV. THE SPECIFIC POSITIONING OF ACYL CHAINS DURING *DE NOVO* PHOSPHOLIPID BIOSYNTHESIS

There is a widely held conviction that the enzymes catalyzing the net synthesis of glycerophospholipids from fatty acids and water-soluble precursors are at least partly responsible for the highly nonrandom fatty acid placement in the final products. This viewpoint is supported by the finding that all phospholipid precursors from phosphatidic acid on do contain mainly saturated fatty acids in their *sn*-1 position and unsaturated fatty acids in their *sn*-2 position.

Establishing just how this specific positioning is achieved has been far from easy. The addition of the first acyl group to *sn*-glycerol 3-phosphate is catalyzed by the enzyme acyl-CoA:*sn*-glycerol 3-phosphate acyltransferase (EC 2.3.1.15), which is found both in the microsomes and the mitochondria of animal cells and in the chloroplast envelope of plants. Bell and Coleman[64] have reviewed recent work on the activity of this enzyme. The relative activities of the microsomal and mitochondrial enzymes vary from tissue to tissue and also during cell differentiation, making general statements regarding the comparative importance of the two activities meaningless. The two enzymes differ in a number of ways, including their specificity for fatty acyl-CoA.[65] While the mitochondrial acyltransferase strongly prefers saturated acyl chains, that associated with microsomes showed little selectivity between palmitoyl-CoA and oleoyl-CoA.

A second fatty acyl group is added to lysophosphatidic acid by the microsomal enzyme acyl-CoA:1-acyl-*sn*-glycerol 3-phosphate acyltransferase (EC 2.3.1.51). This enzyme forms phosphatidic acid having mainly unsaturated fatty acids in the *sn*-2 position, at least in cells active primarily in synthesizing phospholipids rather than triglycerides.[64]

Phosphatidic acid can also arise via two other pathways, which are probably of minor quantitative significance in most tissues. These are (1) the phosphorylation of 1,2-diacylglycerol by the diacylglycerol kinase activity located in the microsomal and cytosolic compartments of the cell, and (2) the acylation of 1-acyl-glycerol 3-phosphate arising through the reduction of 1-acyl-dihydroxyacetone phosphate.[66] In some tissues, e.g., mammalian heart and brain, acyldihydroxyacetone phosphate also serves as the substrate for the synthesis of ether lipids. In this the acyl group is exchanged for a long chain alcohol, which becomes attached to the *sn*-1 position of the glycerol moiety by an ether linkage.[66]

Phosphatidic acid serves as a branch point in *de novo* phospholipid biogenesis. It is either dephosphorylated by phosphatidic acid phosphatase (EC 3.1.3.4), yielding diacylglycerol, or it is converted to CDP-diacylglycerol by the enzyme CTP:diacylglycerolphosphate cytidyltransferase. Diacylglycerol generated by the former reaction is utilized for the synthesis of phosphatidylcholine, phosphatidylethanolamine, or triglyceride, whereas CDP-diacylglycerol is the precursor of phosphatidylserine, phosphatidylinositol, phosphatidylglycerol, and cardiolipin (Figure 5).

A great deal of uncertainty still surrounds the issue of whether some or all of the above enzymes exhibit an in vivo preference for substrates having a specific acyl chain composition. Because of the multiplicity of competing reactions clouding the interpretation of in vivo experiments, much of the pertinent research has been carried out using cell-free systems.

As is true of most other studies of complex lipid metabolism in isolated cell fractions, firm conclusions regarding acyl chain specificity are hard to reach. This is due partly to the inherent difficulty in presenting water-insoluble precursors to the appropriate enzymes in a natural way. Using the most satisfactory incubation conditions now available, no striking diacylglycerol specificity was detected during phosphatidylcholine formation by rat lung microsomes.[67]

Furthermore, the inadvertant inhibition of even one step in the multistep *de novo* phospholipid synthetic pathway is enough to prevent formation of the end product, whereas the shorter pathways for deacylation-reacylation are less susceptible to inactivation. Fox and Zilversmit[68] have recently developed an enriched incubation medium supporting a high rate of *de novo* phosphatidylcholine synthesis by rat liver microsomes. Using this system, which unfortunately still does not synthesize the other phospholipids in high yields, it may be possible to assess the relative importance of *de novo* synthesis vs. deacylation-reacylation in determining the final steady-state phospholipid acyl chain distribution.

V. ACYL CHAIN TURNOVER

A. Phospholipid Deacylation

Most cells contain enzymes capable of hydrolyzing the fatty acids from phospholipids. The enzymes usually display positional specificity; those designated phospholipase A_1 (EC 3.1.1.32) cleave the fatty acid from the *sn*-1 position of phospholipids while the phospholipase A_2 (EC 3.1.1.4) type deacylates phospholipids at the *sn*-2 position (Figure 6). While much of the information available on the properties and reaction mechanism of phospholipases A comes from studies of the more easily purified extracellular enzymes found in venoms and fungal secretions, our discussion will be restricted to enzymes residing within the cell and acting upon membrane structural lipids as a part of normal metabolism. We will not consider the poorly characterized lysophospholipases (sometimes termed phospholipases B), which remove the remaining acyl group from lysophospholipids.

Also not discussed in this section are the deacylating enzymes occurring in higher plants. These hydrolases show little positional specificity and attack both phospholipids and glycolipids in scenescing or physically disrupted higher plant tissue.[69] They are presumably involved in the in vivo phospholipid fatty acyl turnover known to take place in growing plants, but little detailed information is available concerning their mode of action.

1. Phospholipase A_2

The biochemistry of both the intracellular phospholipases A has been thoroughly reviewed by Van den Bosch.[70] He described a wide variety of bacterial and animal tissues as containing both enzyme activities, leading one to suspect that the enzymes are virtually ubiquitous in these organisms. There are some apparent differences in the subcellular localization of the two enzymes. These differences are best authenticated in rat liver, which has been studied

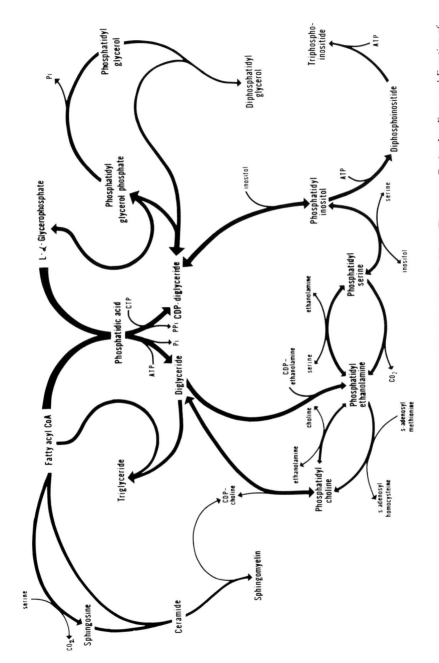

FIGURE 5. Summary outline of phospholipid metabolism interrelationships. (Modified from Thompson, G. A., Jr., *Form and Function of Phospholipids*, Ansell, G. B., Dawson, R. M. C., and Hawthorne, J. N., Eds., Elsevier, Amsterdam, 1976, 67.)

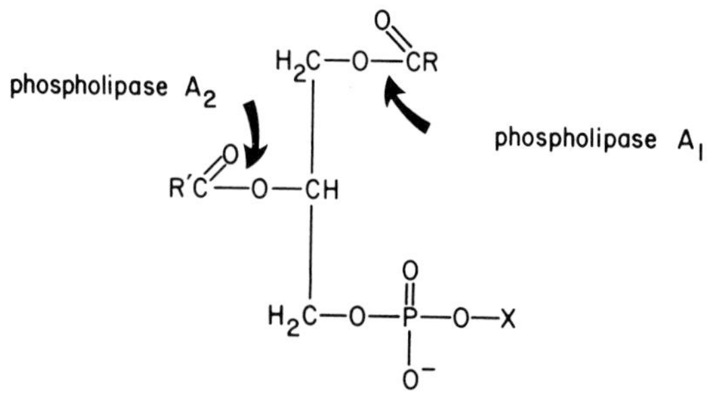

$$H_2C\!-\!O\!-\!\overset{\displaystyle O}{\overset{\|}{C}}R$$

phospholipase A_2

$$R'\overset{\displaystyle O}{\overset{\|}{C}}\!-\!O\!-\!CH$$

phospholipase A_1

$$H_2C\!-\!O\!-\!\overset{\displaystyle O}{\underset{\underset{O^-}{|}}{\overset{\|}{P}}}\!-\!O\!-\!X$$

FIGURE 6.   The sites of phospholipase A_1 and phospholipase A_2 attack.

in several laboratories. Thus, phospholipase A_2 has been found associated with various organelles, including the plasma membrane, mitochondria, Golgi apparatus, lysosomes, and cytosol.[70] With the exception of the lysosomal enzyme, which is released in a soluble form,[71] phospholipase A_2 of formed organelles is membrane-associated. It is noteworthy that liver endoplasmic reticulum, while enriched in phospholipase A_1, contains only trace amounts of phospholipase A_2.[70,72]

Rat lung microsomes, on the other hand, exhibited phospholipase A_2 activity but no phospholipase A_1 activity. The phospholipase of these preparations, incubated in the absence of detergents, showed a greater preference for endogenous phosphatidylcholine species containing an unsaturated fatty acid at the *sn*-2 position than for disaturated species.[73]

Phospholipase A_2 may also show a specificity among the unsaturated fatty acids normally located at the *sn*-2 position of natural phospholipids. Thus Waite and Sisson[74] found a phospholipase A_2 purified from an acetone powder of rat liver mitochondria to release oleic, linoleic, linolenic, and arachidonic acids, in that order of preference, from liver phosphatidylethanolamine. The tendency to cleave the unsaturated acyl group having the least number of double bonds was not influenced by the nature of the fatty acid at the *sn*-1 position.

Most membrane-bound phospholipases A_2 which have been purified seem to have molecular weights of 12,000 to 14,000.[70] With the exception of the pancreatic enzyme, they have not been found to occur in a zymogen form. Although enzymatic activity appears to be stimulated or inhibited by enzyme-associated proteins in some cases,[70] the most widespread mode of regulation in vitro is through the availability of Ca^{2+}. It is tempting to postulate a regulatory role for Ca^{2+} in vivo. The intracellular concentration of Ca^{2+} can be regulated by entry of the cation from outside the cell or release into the cytoplasm from intracellular stores, sometimes mediated through the hormone-induced production of cyclic AMP.

The finding that calmodulin stimulates the phospholipase A_2 activity of human platelets,[75] suggests that the mechanism of Ca^{2+} activation may involve a Ca-calmodulin complex. Mouse peritoneal macrophages have been shown to contain a phospholipase A_2 whose activity appears to be greatly stimulated in vitro by the action of Ca^{2+}-requiring protein kinase.[76] Direct evidence for phosphorylation of the enzyme is not yet available. Despite these many indications of Ca^{2+} involvement, it still has not been established that phospholipase activity in the intact cell is ever limited by the availability of Ca^{2+}.

As with other lipid metabolic enzymes, correlating activities of purified phospholipases has been confounded by the hydrophobic nature of the enzymes and their substrates. Consequently, measurements of phospholipase activity, including some of those described above, have often involved the use of detergents. Although generally undesirable because of the tendency to disrupt normal membrane structure and alter enzyme specificity, the addition of detergents has frequently been deemed essential in order to bring exogenous substrates into contact with the enzyme or to recover measurable amounts of product.

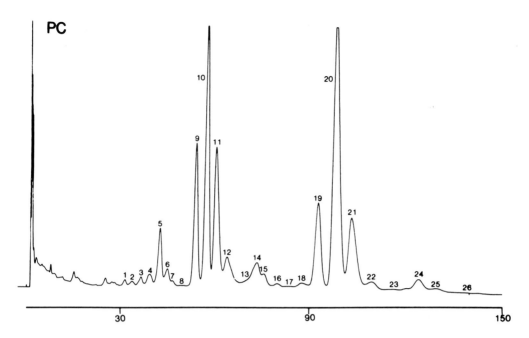

FIGURE 12. High-performance liquid chromatographic separation of rat liver phosphatidylcholine molecular species.[106] For details of the separations, which are based on chain length and degree of unsaturation, refer to the original reference. (From Patton, G. M., Fasulo, J. M., and Robins, S. J., *J. Lipid Res.*, 23, 190, 1982. With permission.)

thetase-deficient mutants of *E. coli.*[95] A second contrast in bacterial cells is the low level of phospholipid acyl chain turnover mentioned earlier. Changes in molecular species generally involve the complete disassembly of certain phospholipids and their replacement by freshly synthesized ones.

C. Evidence for Phospholipid Deacylation-Reacylation as a Mechanism for Membrane Fluidity Alteration

Of all the enzymatic mechanisms having the potential for changing membrane physical properties, this is probably the least thoroughly investigated. One problem has been the technical difficulties of detecting and quantifying deacylation-reacylation produced changes in lipid structure. Only within the past year or so has it become possible to clearly resolve phospholipid molecular species or their derivatives chromatographically. Recent advances in high performance liquid chromatography[106] (Figure 12) and coupled gas chromatography-mass spectrometry[107] confirm the feasibility of routine analysis of this type.

The other problem in evaluating the role of molecular species alteration per se is that such alterations seldom occur in the absence of other structural changes. Net changes in fatty acid unsaturation, phospholipid polar head group distribution, etc., as discussed earlier in this chapter, usually overshadow any effects caused solely by inter- and intramolecular rearrangement of preexisting fatty acids. However, in a few instances, phospholipid molecular species changes in the absence of other detectable lipid modifications have suggested that they can be very important physiologically.

One intriguing case is that involving a mutant strain of *Drosophila melanogaster* resistant to ether and a number of other volatile anesthetics.[108] Phospholipids of the resistant flies were shown to differ from those of the sensitive wild type only in having an increase in the phosphatidylethanolamine molecular species containing 34 fatty acid carbons and 3 double bonds, and an accompanying decrease in phosphatidylethanolamine species having carbon numbers of 36:2, 36:3, and 36:5. The direct involvement of this simple compositional change in the insect's resistance to anesthetics must for the moment be left open pending further study.

In another case, the correlation between lipid molecular species changes and membrane fluidity change has some experimental support. Following sudden chilling from 39°C to 15°C, the microsomal phospholipids of *T. pyriformis* cells underwent a substantial reorganization of their molecular species within 1 hr, well before a significant increase in fatty acid unsaturation or other lipid changes were detectable.[109] These altered fatty acyl pairings reflected a remarkable low temperature-induced change in acyltransferase and/or phospholipase specificity which caused the replacement of saturated fatty acids normally present at the phospholipid *sn*-1 position with γ-linolenic acid.[110] During this initial period when only phospholipid molecular species were being modified, the fluidity of the membrane, as monitored through fluorescence depolarization measurements, steadily increased towards that of cells fully acclimated to low temperature.[111]

An even more clear-cut instance of isolated molecular species rearrangement was observed in *Tetrahymena* ciliary membranes. Although these organelles are physically and metabolically remote from the centers of lipid formation and turnover in the cells, they were found to sustain a significant revamping of their many phospholipid molecular species well before any measurable importation of lipids from other cellular compartments could take place.[102,112] Since this may indeed be the only type of rapid *in situ* modification possible in the cilia, it may provide enhanced short-term stability to low temperature stress until the more protracted movement of lipids into cilia from other cell parts can be completed.

Unfortunately, it is not yet possible to interpret molecular changes such as those described above in terms of their detailed effects on membrane physical properties. Many more empirical studies of simple two and three component systems, especially those containing polyunsaturated fatty acids, will be necessary before the physical behavior of complex natural mixtures can be understood.

VI. OTHER MODIFICATIONS AFFECTING PHOSPHOLIPID ACYL CHAIN COMPOSITION

Apart from the fatty acids incorporated into phospholipids during their *de novo* synthesis and those inserted as replacements by deacylation-reacylation reactions, there are numerous other reactions by which the fatty acid composition of a particular phospholipid can be affected. There can be quantitatively significant transfers of entire diglyceride moieties from one phospholipid class to another by any one of several well-known pathways, some of which are outlined in Figure 5. For example, approximately 20% of the molecular species of phosphatidylcholine in mammalian liver[113] is recruited from the phosphatidylethanolamine pool by sequential methylation of the latter phospholipid. Although the quantitative importance of the methylation pathway is much less in nonhepatic tissues of mammals, it is a major source of phosphatidylcholine in many simpler organisms. A consideration of this pathway and its effect on membrane fluidity is presented in Chapter 6.

In addition, base exchange reactions can lead to a direct transfer of diglyceride moieties from one phospholipid class to another. The relative importance of this pathway as a contributor to the steady-state molecular species composition of tissue phospholipids is not known.

Many other examples could be chosen. Phosphatidylcholine is believed to be the principal donor of linoleic acid incorporated into monogalactosyl diglyceride by plants.[114] A full discussion of this and the other potentially regulated means by which fatty acids can indirectly find their way into a particular membrane lipid class is beyond the scope of this chapter. However, indirect entry of fatty acyl groups into a certain lipid may in some cases be of greater consequence than the more general routes of incorporation we have emphasized here.

However, some information might be derived from data available for ceramides, cerebrosides,[68,71,75] and model compounds of SPM.[76]

Most of the physical studies on SPM were with liposomal dispersions which served as models for biological membranes. These were in the form of large multilamellar liposomes (see Figure 1) or in the form of unilamellar vesicles. It is worth noting that the properties of the liposomes depend on their composition and their curvature. The effect of liposomes curvature on their properties was clearly demonstrated for PC.[77-79] In addition, many of the physical properties of SPM, as with PC, might vary with the acyl chain composition. This is true for such properties as thermotropic behavior,[80,81] osmotic properties of liposomes,[58] and liposomal size.[67,82] Thus, specifying the origin of natural SPM may not suffice, since the composition of SPMs of identical source, such as bovine brain, differs for different preparations[81] (Table 3). This might stem from a dietary effect[37,63,64] or from variations in the purification procedures.[229]

C. Monomolecular Films

The values of the limiting surface area per molecule, obtained from monolayer studies, are in agreement with the corresponding values obtained from X-ray diffraction studies of lipid bilayers. The following average values of limiting surface area were reported; dipalmitoyl PC-44.5 Å^2; dioleoyl PC-72 Å^2; egg PC-62 Å^2; bovine brain SPM — 42 to 50 Å^2 (the exact value depends on the acyl chain composition). The limiting surface area of synthetic sphingomyelin (racemic, D,L-*erythro*) is similar to that of semisynthetic PCs; C16 SPM - 40.5 Å^2; C18 SPM - 39 Å^2; C24 SPM - 38 Å^2.[83] Similar results were previously described for the natural D-*erythro* SPM, isolated from bovine brain which was enriched with *N*-stearoyl-sphingosylphosphorylcholine. The liquid-condensed film had a minimal molecular surface area of only 40 Å^2 while a fraction enriched with *N*-nervonyl-sphingosylphosphorylcholine had a corresponding value of 56 Å^2, and a film of the liquid-expanded type.[84] The difference in the surface properties of these two fractions is most probably related to the *cis* double bond between carbon atoms 15 and 16 in the nervonyl residue, and not to the *trans* double bond between carbon atoms 4 and 5 of the sphingosyl residue.[84-86] It is worth noting that this *trans* double bond has a considerable effect on the force/area curves of ceramides. Thus, *N*-octadecanoyl-D-dihydrosphingosine, which has a limiting area similar to that of *N*-octadecanoyl-D-sphingosine (42 Å^2), is much more expanded at lower pressure. This is due to facilitations of chain stacking and close packing of the lipid molecules, thereby imposing a more condensed organization of the lipid.[73] Another possible role of the *trans* double bond of the sphingosine base was derived from measurements of the surface electrical potential of films of dipalmitoyl PC and bovine brain SPM. Although measurements were made under similar conditions, the values were markedly different. The relatively larger surface potential of SPM monolayers was reduced by hydrogenation, suggesting a contribution of the *trans* double bond to the surface potential.[86]

The three synthetic D,L-*erythro* SPMs (C16, C18; C24 SPM) resemble the disaturated PCs. They exhibit a discontinuity in their force/area curves which display transition from liquid-condensed to liquid-expanded state.[87-90] Although the nature of this transition is a controversial issue,[91-93] it has a major effect on the rate of penetration and hydrolysis of the monomolecular layer of these SPMs by *Staphylococcus aureus* sphingomyelinase which is optimal at the transition range.[83]

It is also worth demonstrating the analogy of phase behavior between SPM organized as a monomolecular layer at the air-water interface and as lipid bilayers. For both C16 and C24 SPMs, this transition occurs at greater surface pressure than for C18 SPMs. Thus, below 10 dyn/cm the films of C16 and C24 SPM are more expanded than those of C18 SPM.[83] This resembles the thermotropic behavior pattern of these three SPMs[80] and will be discussed below in more detail (Section III.E.1).

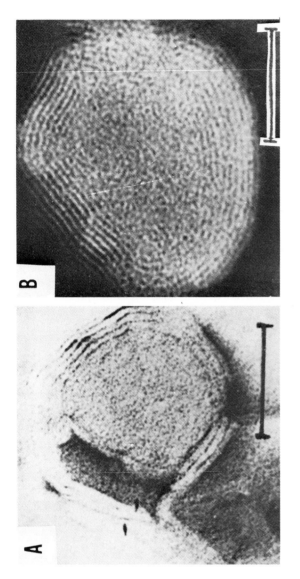

FIGURE 1.   Electron micrograph of multibilayered liposomes made of bovine spinal cord SPM containing 10% dicetylphosphate (A) and of egg PC containing 10% dicetylphosphate (B). The final magnification was 159,000 for both. A 1000 Å measure is shown on both. (From Hertz, R. and Barenholz, Y., *Chem. Phys. Lipids*, 15, 138, 1975. With permission.)

D. Solubility in Organic Solvents

Considerable differences in solubility in organic solvents were found between SPM and PC. Thus, in nonpolar solvents (such as decane which is used for preparation of planar lipid bilayers), SPMs, but not PCs, form a gel.[94] Naturally occurring SPMs and most synthetic SPMs are practically insoluble in chloroform even at high temperatures, while most PCs form micellar solutions in this solvent.[95-97] Solubilization of SPM as micelles in chloroform-based solvents can be increased by addition of a polar solvent such as methanol. Even then, considerable differences are observed in the solubility of SPM and PC. Alpes[94] suggested that the solubility of SPM can be further increased and made comparable to that of PC by adding water. It is plausible that the differences in the solubility in organic solvents is caused mostly by differences in the interfacial region of these two phospholipids.

E. Thermotropic Behavior

The melting point of natural SPM (of unspecified origin) to an isotropic liquid occurs at 210°C.[72] The melting points of synthetic sphingomyelins are independent of the length of the acyl chains which is also true for PC and PE.[72] Unsaturated fatty acids reduce the melting point; this is related to the location of double bonds and possibly to *cis* or *trans* isomerism. Also, the double bond in the sphingosine affects this property, since dihydrosphingomyelin has a higher melting point than SPM. As expected there is almost no difference in melting point between the various stereoisomers of the same SPM.

Similar to other phospholipids, phase changes other than melting to an isotropic liquid occur at lower temperatures.[12,72] In a pioneering study of the thermotropic and lyotropic behavior of lipids using a low angle X-ray diffraction, Reiss-Husson[98] found that an aqueous dispersion of bovine brain SPM forms a lamellar gel phase at 25°C, but a liquid crystalline lamellar phase at 40°C. At 40°C the lamellar phase incorporates a maximum of 40% water (by weight) with additional water forming a bulk phase. Reiss-Husson[98] also found that the surface area per molecule, as well as the interlamellar spacing, increases as SPM is progressively hydrated (swelling process). At the maximum hydration of 40%, the area per molecule was found to be 54 Å^2, and the lipid bilayer and water layer thicknesses were 40 and 30 Å, respectively. It is worth noting that in many cases, SPM, which is described as "anhydrous", exists at least partially in hydrated form, as was concluded from infrared spectroscopy.[53]

The phase behavior of aqueous dispersions of SPM having a well-defined fatty acid composition was studied by Shipley and co-workers[99] using polarized light microscopy, differential scanning calorimetry, and X-ray diffraction. Lamellar phases, in which water intercalated between sheaths of lipid molecules arranged as bilayers, were found to be present over much of the phase diagram. An order-to-disorder transition separates the high temperature liquid crystalline lamellar phase from a more ordered lamellar phase (gel phase) at low temperature. The thermotropic behavior in the absence of water proved to be similar to that exhibited by various PCs. At 87°C, a transition occurs from a crystalline phase, in which SPM is organized in bilayers, to a liquid crystal-like phase with a lamellar structure. Formation of a viscous isotropic phase occurs at 144°C and, at 170°C, is transformed to give a hexagonal-type structure with the lipid headgroups forming a core of parallel rods packed in a regular two-dimensional hexagonal lattice. Increase of water content causes the gel-liquid crystalline transition temperature to decrease progressively to a value of about 40°C at 35% water which is then independent of further increase in water content. At 47°C, this sphingomyelin preparation shows a maximum water uptake of 35% (w/w). Above this water content, the maximally swollen lamellar lipid phase coexists with an excess bulk water phase. At this limiting hydration, the area per molecule was found to be 60 Å^2 and the lipid bilayer and water layer thicknesses were 38 and 22.2 Å, respectively. The differences between these values and those reported earlier by Reiss-Husson[98] may be due to differences in fatty acid composition of the SPM but are more likely due to differences in the temperature of

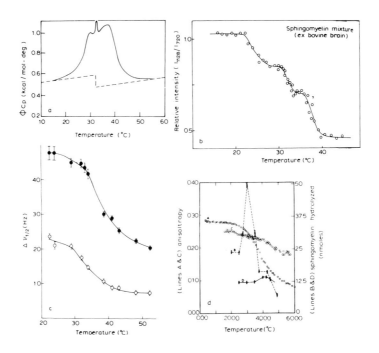

FIGURE 2.   Thermotropic behavior of bovine brain SPM-liposomes: (a) a calorimetric scan of MLV relating °C_p and temperature; (b) "melting curve" of
SPM-MLV derived from the change in the *trans* bond intensity relative to the
temperature invariant C-N stretch; (c) effect (Raman spectroscopy) of temperature
on the linewidths of methylene (●) and choline-methyl (○) proton resonances
using small unilamellar vesicles (SUV); (d) correlation between the thermotropic
behavior of MLV as measured by fluorescence depolarization of DPH and the
rate of hydrolysis of SPM by the enzyme sphingomyelinase of *S. aureus* (□, ■-
SPM; ○, ●, SPM with 40 mol% cholesterol). For more details see text. (From
Barenholz, Y. and Gatt, S., in *Phospholipids*, Hawthorne, J. N. and Ansell, G.
B., Eds., Elsevier, New York, 1982. With permission.)

the measurements.[99] Shipley and co-workers have pointed out[99] that the maximum hydration
35% observed with bovine brain SPM at 47°C is similar to that of egg PC at 5°C. Since
either system, at the respective temperatures, is just above the gel-liquid crystalline phase
transition, it was suggested that the PC headgroup, which is common to both, is the principal
factor controlling the swelling behavior of the two phospholipids.[99] These X-ray data led
authors to assume a β-type structure in which the hydrocarbon chains are packed in a
pseudohexagonal lattice with rotational disorder,[100,101] similar to PC with heterogeneous acyl
chains. They also suggested that the changes in the molecular packing which occur with
increasing hydration may be due to a progressive tilt of the hydrocarbon chain axis in a β-
type structure; this structure is known to occur in synthetic diacyl PCs.[102] The polymorphic
phase behavior of bovine brain SPM was also confirmed using ³¹P-NMR studies.[103]

Most SPMs isolated from natural sources exhibit a distinct gel-to-liquid crystalline phase
transition. In most cases, the region of the phase transition, when the SPM is present in
excess water, is in the physiological temperature range.[81,99,104,105]

The characterization of the thermotropic behavior of multilamellar large vesicles made of
SPM isolated from bovine brain is described in Figure 2 using physical and biological
methods. From Figure 2 it is clear that the thermotropic behavior of this bovine brain SPM
is rather complicated. A similar conclusion was reached by Calhoun and Shipley[81] for a few
different preparations of bovine brain SPM. In order to understand it better, it is necessary
to have a deeper insight into the thermotropic behavior of SPMs with one defined acyl chain.

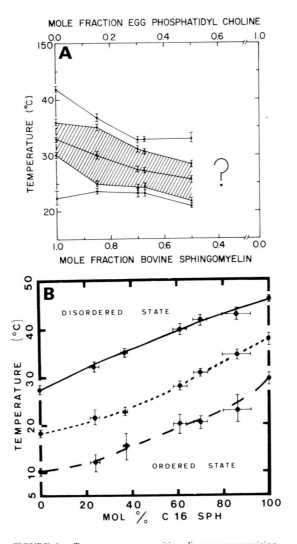

FIGURE 6. Temperature composition diagram summarizing the phase behavior of small unilamellar vesicles (SUV) composed of mixtures of egg PC with bovine brain SPM (left 6A) and for DMPC — D,L-*erythro* C16 SPM (right Figure 6B). The delimiting temperatures were determined from the Arrhenius plots describing the natural logarithm of the apparent microviscosity as function of the reciprocal of the temperature (K^{-1}) as described by Shinitzky and Barenholz (1978). The bars represent uncertainty in break temperatures. For egg PC-bovine brain SPM, the hatched region represents the region in which the major change in "apparent microviscosity" with 1/T was obtained. The question mark is placed where no detectible phase transition or phase separation was obtained using DPH as a probe. (From Barenholz and Thompson, in preparation.) For DMPC — D,L-*erythro* C16 SPM (Figure 6B) the area between the upper line (--) and the middle line (. . .) represents the region in which the fast change in apparent microviscosity occurred while the area between the middle line (. . .) and the lower line (—) represents the region in which the slow change in the apparent microviscosity occurred. (Reprinted with permission from Lentz, B. R., Hoechli, M., and Barenholz, Y., *Biochemistry*, 20, 6803, 1981. Copyright 1981 American Chemical Society.)

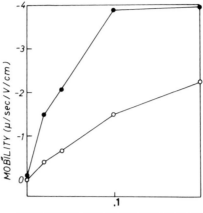

FIGURE 7. The effect of dicetylphosphate to phospholipid mole ratio on the electrophoretic mobility of MLV made of dicetylphosphate-egg PC (●-●) and dicetylphosphate-bovine spinal cord SPM (○-○). (From Shinitzky, M. and Barenholz, Y., *J. Biol. Chem.*, 249, 2652, 1974. With permission.)

The phase diagrams of mixtures of L-dimyristoyl-phosphatidylcholine (DMPC) with D,L-*erythro-N*-palmitoyl-sphingosylphosphorylcholine (C16 SPM) are shown in Figures 10, 11, and 12. Temperature-induced changes in these mixtures were studied using steady-state fluorescence polarization of DPH, as well as freeze-fracture electron microscopy (Figure 12). The phase diagram was almost independent of the type of the vesicles used; therefore, the effect of vesicle curvature on the system's properties is minimal. Also, the anisotropy of DPH fluorescence was found to be almost invariant with C16 SPM content at temperatures just above or below the gel–to–liquid crystalline phase transition.[110] This implies that the acyl chain composition is the main structural parameter which determines the phase separation in systems containing SPM and PC and confirmed the conclusions reached by Calhoun and Shipley.[109]

All the above observations suggest that the main effect of SPMs on biological membrane order and dynamics is through their acyl chains which, in most biological membranes, are more saturated for SPM than for PC or other phospholipids. This difference of degree of saturation probably reflects the different enzymatic pathways involved in the biosynthesis of PC and SPM and enables a metabolic control of the physical properties of the membranes.

B. Interaction of SPM with Cholesterol

Not much attention has been given to the examination of the interactions between cholesterol and sphingolipids. Vandenheuvel[142] suggested, on theoretical grounds, that molecular configuration and van der Waals' interactions should lead to a stable complex between cholesterol and sphingomyelin. Indeed, it has been known for some time that there is a significant positive correlation between the content of cholesterol and sphingomyelin in the membranes of many mammalian cells.[143]

Early ³H-NMR and ESR studies[124,144] showed that, in multilamellar liposomes formed from cholesterol and bovine brain SPM, the effect of the added cholesterol was to "fluidize" the bilayer below the phase transition while making it less fluid above the transition temperature. Thus, the effect of cholesterol on the "apparent microviscosity" of SPM bilayers parallels that observed in glycerophospholipid systems.[72,126,137] ¹H-NMR studies on small unilamellar vesicles composed of cholesterol and SPM showed that in systems containing

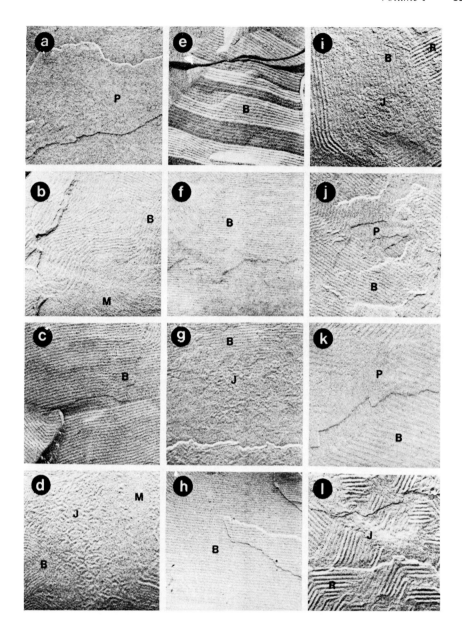

FIGURE 12. Electron micrographs of freeze–fracture replicas prepared from DMPC/D,L-*erythro* C16 SPM MLV quenched from several temperatures. Magnification 75,000 x. Samples shown are the following: pure C16 SPM quenched from (A) 55°C, (B) 43°C; (C) 34.5°C; (D) 30°C; (E) 20°C; (F) pure DMPC from 18°C; (G) 23 mol% C16 SPM from 28°C; (H) 46 mol% C16 SPM from 22°C; (I) 92 mol% C16 SPM from 29°C; (J) 2 mol% C16 SPM 5°C; (K) 10 mol% C16 SPM from 7°C; (L) 92 mol % C16 SPM from 13°C. Fractured samples were platinum shadowed at an angle of 45°. The micrographs are shown with the shadowing from below. (Reprinted with permission from Lentz, B. R., Hoechli, M., and Barenholz, Y., *Biochemistry*, 20, 6803, 1981. Copyright 1981 American Chemical Society.)

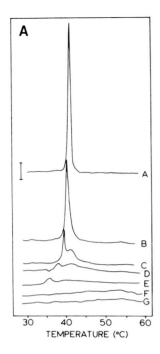

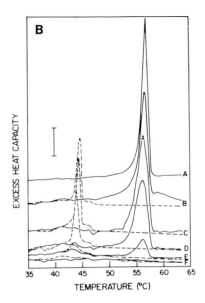

FIGURE 13A.   C16 SPM: samples
contained (A) 0.0; (B) 6.2; (C) 11.6;
(D) 15.6; (E) 19.3; (F) 26.0; (G) 30.3
mol% cholesterol.

FIGURE 13B.   C18 SPM: Scans performed after at
least 7 days at room temperature are shown as solid
lines. Dashed lines represent scans performed immediately after samples were cooled to room temperatures.
The vertical bar represents 2 kcal mol^{-1} deg^{-1}. Samples
contained (A) 0.0, (B) 1.1, (C) 5.1, (D) 11.0, (E) 16.8,
and (F) 20.0 mol% cholesterol.

FIGURE 13.   The thermotropic behavior of MLV made of mixtures of cholesterol with D.L-*erythro*
C16 SPM (Figure 13A); D.L-*erythro* C18 SPM (Figure 13B) and D.L-*erythro* C24 SPM (Figure 13C).
Phospholipid concentration was approximately 10 mM in all samples. Heat-capacity values were
calculated per mole of phospholipid. The vertical bar represents 1 kcal mol^{-1} deg^{-1}. (Figures 13A
and 13C reprinted with permission from Estep, T. N., Mountcastle, D. B., Barenholz, Y., Biltonen,
R. L., and Thompson, T. E., *Biochemistry*, 18, 2112, 1979. Copyright 1979 American Chemical
Society; Figure 13B reprinted with permission from Estep, T. N., Freire, E., Anthony, F., Barenholz,
Y., Biltonen, R. L., and Thompson, T. E., *Biochemistry*, 20, 7115, 1981. Copyright 1981 American
Chemical Society.)

the one found for the preferential interaction with cholesterol as measured by calorimetry.[148,149]

Interaction of cholesterol oxidase with membranous cholesterol can serve for equivalent
purposes. This enzyme oxidases cholesterol to give cholest-4-en-3-one. For many intact
biological membranes and for intact lipoproteins, the cholesterol is not available for oxidation
by this enzyme. These include membranes of red blood cells,[24,153,154] viruses,[24] and mycoplasma.[241] Similar results on the lack of availability of cholesterol to oxidation by cho-
lesterol oxidases were obtained for small vesicles made of mixtures of PC-cholesterol, SPM-
cholesterol, PC-PE-cholesterol, and PC-SPM-cholesterol when the cholesterol content was
lower than 42 to 45 mol%.[24,155] For the above mixtures the cholesterol oxidation was not
increased below, in, or above the gel-to-liquid crystalline phase transition of the phospholipid.[155] Almost 100% accessibility of cholesterol was obtained for PS-cholesterol and PS-
PE-cholesterol mixtures. Of special interest is the observation that a large increase in cholesterol availability in either PC or SPM bilayers occurred when the cholesterol levels
exceeded 45 mol%.[24,155] Under such conditions all cholesterol is oxidizable in the case of
PC-cholesterol mixtures, but only 60% of the cholesterol is oxidizable in SPM-cholesterol
mixtures. This also supports the notion of stronger SPM-cholesterol interactions when compared with that of PC.

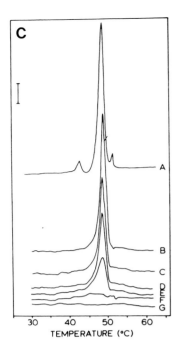

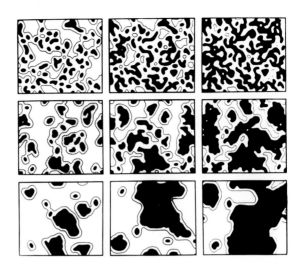

FIGURE 13C.   C24 SPM: The vertical bar represents 1 kcal mol^{-1} deg^{-1}. Samples contained: (A) 0.0, (B) 6.9, (C) 12.4, (D) 15.0, (E) 20.1, (F) 24.4, and (G) 29.9 mol% cholesterol.

FIGURE 14.   Diagrammatic illustrations of typical computer simulations for the lateral distribution of lipid species in binary mixtures of cholesterol and D.L-*erythro* C16 SPM (top row), L-DPPC (middle row), and D.L-*erythro* C24 SPM (bottom row) at 8, 15, and 22 mol% cholesterol (left, center, and right columns, respectively). The black areas represent cholesterol-rich domains, the gray areas represent free lipid domains, and the stippled areas represent regions of boundary lipid. These figures were obtained by using periodic 60 × 60 cholesterol-lipid lattices. (From Snyder, B. and Freire, E., *Proc. Natl. Acad. Sci. U.S.A.*, 77, 4055, 1980. With permission.)

V. INTERACTION OF SPHINGOMYELIN WITH PROTEINS

A. Relation to Enzyme Specificity and Activity

Although the phosphorylcholine moieties of SPM and PC are chemically identical, most enzymes which hydrolyze this group, namely phospholipase C and sphingomyelinase, are specific for SPM or PC.[156-166] For sphingomyelinase from *S. aureus*, the enzymatic activity could be related to the thermotropic behavior of the multilamellar vesicles (Figure 2D). Optimal activity appears at the phase transition region and depression of this transition by cholesterol also depresses the enzymatic activity.[167] Enzymatic hydrolysis by the above enzyme of monomolecular films of synthetic and natural SPMs is also optimal at the transition from the liquid-condensed to the liquid-expanded states.[83] This is in accordance with the hydrolysis of PC by pancreatic phospholipase A$_2$ which is also enhanced by a phase separation, imposed by the presence of SPM.[234,235]

B. Specific Interactions of SPM with Proteins

Specific interaction of SPM with membrane proteins other than sphingomyelinases, has also been proposed. Kramer and co-workers[168] showed, in reconstitution experiments, that the proteins of the sheep erythrocytes bind more strongly to SPM than the corresponding proteins of the human erythrocyte. This preferential binding correlates with the unusually high SPM content of the sheep erythrocyte membrane shown in Table 4. Widnell and co-workers[169] reported that the 5′-nucleotidase from the plasma membranes of rat liver cells is isolated as a complex with SPM; about 100 mol of lipid are found per mole of protein. Removal of sphingomyelin inactivates the enzyme.[169,170] In comparison, Evans and Gurd[171] purified a similar enzyme from mouse liver plasma membranes, which was active in the absence of any lipid. Sandermann[49] and McElhaney[172] discussed in detail the problems of

Table 4
PHOSPHOLIPID DISTRIBUTION IN ERYTHROCYTES FROM VARIOUS MAMMALIAN SPECIES[a]

	Dog	Rat	Guinea pig	Horse	Rabbit	Human	Cat	Pig	Sheep	Cow	Goat
Choline phospholipids											
PC	46.9	47.5	41.1	42.4	33.9	35.7	30.5	23.3	Traces	Traces	Traces
SPM	10.8	12.8	11.1	13.5	19.0	24.7	26.1	26.5	51.0	46.2	45.9
LPC	1.8	3.8	0.3	1.7	0.3	n.r.	0.3	0.9	n.d.	n.d.	n.d.
Amino phospholipids											
PE	22.4	21.5	24.6	24.3	31.9	24.7	22.2	29.7	26.2	29.1	27.9
PS	15.4	10.8	16.8	18.0	12.2	13.8	13.2	17.8	14.1	19.3	20.8
Acidic phospholipids											
PA	0.5	0.3	4.2	0.3	1.6	n.r.	0.8	0.3	0.3	0.3	0.3
PI	2.2	3.5	2.4	0.3	1.6	5.8	7.4	1.8	2.9	3.7	4.6
Others									4.8	1.7	0.8
SPM/PC	0.23	0.27	0.27	0.32	0.56	0.64	0.86	1.13	>12.0	>12.0	>12.0

Note: PC, phosphatidylcholine; PE, phosphatidylethanolamine; PS, phosphatidylserine; PI, phosphatidylinositol; PA, phosphatidic acid; SPM, sphingomyelin; LPC, lysophosphatidylcholine; n.d., not determined; n.r. not recorded.

[a] From References 9 and 10

establishing specific lipid requirements for membrane-bound enzymes. The (Na^+-K^+)-ATPase isolated from rabbit kidney by Lubrol® extraction binds strongly to liposomes made of SPM but not to dispersions of PC unless the positively charged amphiphile, stearylamine, is added.[173] Recently, it was shown that the hemolytic toxin isolated from the sea anemone, *Stoichactis heliauthus*, binds to aqueous dispersions of SPM but not to liposomes formed from other erythrocyte lipids. This observation suggests that the site of binding of this cytolytic glycoprotein might be SPM of the plasma membranes.[174]

The interaction between SPM and acetylcholinesterase(AchE)[175] is very interesting. Using a flotation-type assay, the purified AchE was found to bind strongly to liposomes composed of bovine brain SPM but not of egg PC.[175] This enzyme also binds well to liposomes made of synthetic D,L-*erythro* SPM.[117] The binding to SPM liposomes was unaffected by the thermotropic behavior of the SPM by the liposome curvature or the stereospecificity of SPM, but it was extensively reduced by the presence of PC in the bilayer. The binding of AchE did not damage the integrity of the liposomal membrane; it also had no effect on the kinetic parameters of the enzyme substrate interaction. The only effect which was found was a major increase of enzyme stability when present in dilute solution.[117]

SPM is also bound to circulating plasma lipoproteins of higher animals. In humans, the ratio of PC to SPM is about 3:1 in the total plasma lipoprotein fraction.[176] The ^{31}P-NMR resonance arising from each of these phospholipids is similar to that observed in vesicles formed from mixtures of these phospholipids.[112,141] The results suggest that the interactions between these phospholipids and the apolipoproteins do not involve the phosphorylcholine moiety. In reconstitution experiments, however, SPM appears to bind strongly to apolipoprotein A-II but not to A-I, which interacts preferentially with PC.[177]

The molecular specificities of SPM and PC are sufficient to interact differently with antibodies. Arnon and Teitelbaum[178] showed that antibodies prepared against sphingosylphosphorylcholine bound to a carrier through its free amino group at C2 (see Scheme 3) interacted with membranes enriched in SPM (such as sheep red blood cells; see Table 4), but not with other membranes whose SPM content is low and is replaced by PC, such as guinea pig red blood cells. This suggested that the interface region of the choline phospholipids is responsible for the antibody specificity.[178]

VI. SPHINGOMYELIN IN BIOLOGICAL MEMBRANES

A. Distribution

SPM is a major lipid component of the cellular membranes,[9,10] as well as of the serum lipoproteins[179] of mammals. The content of SPM varies considerably in membranes from diverse sources, but in many systems, the sum of the two choline-containing lipids, SPM and PC, constitutes about half of the total phospholipid although the mole ratio of these two respective components varies considerably.[9,10,12] This is true even for membranes of different organs and tissues of a single species. Thus, in most organs, the PC to SPM ratio is fairly constant in the same organ of different mammalian species.[9] This is not true for brain, where large variations in this ratio between various species are found.[11] In brain, lipid-rich regions lipid-rich regions have higher SPM to PC ratios than do lipid-poor regions. In spinal chord, this ratio is even higher than in the brain.[11]

There is also a distinct pattern of SPM content in various parts of the eye. All the hard tissues in the bovine eye are very rich in SPM.[11] This is in contrast with very low levels of SPM (less than 5% of total phospholipids) in the retina, the rod outer segment, and the neurons.[11]

In the erythrocyte membrane, which was extensively studied, the SPM to PC ratio varies with mammalian species from 0.25 in rats to above 12 in ruminants. Data for 11 species are summarized in Table 4. Cells which have subcellular organelles show the highest SPM to PC ratio in the plasma membrane and the lowest in the nuclear and mitochondrial membranes. The endoplasmic reticulum and Golgi membranes have intermediate values.[9-11] There appears to be an increasing gradient in the SPM to PC ratio from the cell center to the periphery. It is worth noting that most envelope viruses contain relatively high levels of SPM in their membrane which is a reflection of the similarity in lipid composition between the viral membrane and the plasma membrane of the host cell from which it is derived.[24,180-183]

B. Membrane Asymmetry

Considerable evidence suggests that there is a marked asymmetric distribution of lipids between the outer and inner faces of plasma membranes[20-22,184,185] which is best documented in the membrane of the human erythrocyte. Experiments utilizing phospholipases,[135,186-188] phospholipid exchange proteins,[18,25,189] or chemical labeling reagents[190-192] clearly show that essentially all of the SPMs and most of the PCs are located in the external surface of this membrane, while PE and PS are the major lipid constituents of the cytoplasmic surface. Associated with this transmembrane lipid asymmetry is an absolute compositional asymmetry of protein components.[20-22] A similar asymmetric distribution of SPM has been demonstrated in rat erythrocytes,[193] avian erythrocytes,[194] LM cell plasma membranes,[195] and vesicular stomatitis virus.[24,181]

C. Changes in SPM Content Associated with Aging and Pathological Conditions

The aging of tissues which have low rates of phospholipid turnover is followed by major changes in lipid composition of their cell membranes. A good example is the aging of the aorta and arteries in humans, during which there is a marked increase in the relative contents of SPM and cholesterol in the membranes of cells comprising these tissues.[9,196,197] Rouser and Solomon[197] demonstrated a linear correlation between the logarithm of the age (in years) to the total phospholipids, mainly due to a large increase in SPM. Eisenberg et al.[196] showed a similar correlation, expressed as increased SPM to PC mole ratio with age. A more pronounced increase in SPM occurs during the development of atherosclerosis.[198,199] During aging, SPM reaches 40% of the total lipid of the intima and in advanced aortic lesions, it may reach levels as high as 70 to 80% of the total phospholipids. The increased concentration of lipids in the intima, most of which can be attributed to an increase in sphingomyelin, might be a consequence of changes in enzymatic activities and of the marked increase in the entry rate of the serum sphingomyelin into the aortic wall.[200,201]

Eisenberg and co-workers[199] showed a greater incorporation of choline into the phospholipids of the aorta with increasing age, which parallels the increase of phospholipase A activity through sphingomyelinase activity, remains constant, or even declines. These age-dependent changes in enzymatic activities involved in lipid metabolism are all supporting the increase in SPM to PC ratio during aging. It was proposed that increased biosynthesis or intake of the phospholipids is followed by an increased rate of hydrolysis of PC but not of SPM.

SPM accumulation with age was also observed in the agranular endoplasmic reticulum, the plasmalemma of smooth muscle cells, the principal cells of the intima, and the intermedia of the wall.[202-204]

Bierman and co-workers[205] found that rat aortic smooth muscle cells take up ''remnants'' of very low density lipoprotein (VLDL) formed by lipolysis of the VLDL by lipoprotein lipase. These are rich in cholesterol and SPM,[206] and may be one of the main sources of the increased sphingomyelin of the aortal wall.[201]

Changes in the SPM to PC ratio have been noted in muscular dystrophy[207] and in some malignant diseases. Leukemic cells appear to be deficient in both SPM and cholesterol.[208,209] In a thymus-derived leukemia, the reduction in the SPM and cholesterol content occurs by shedding of rigid plasma membrane portions which are enriched in SPM and cholesterol.[210]

SPM and cholesterol deficiency was also observed in plasma membranes from rat thymic lymphoma cells when compared with normal thymocyte membranes.[211] Conversely, it was shown that the SPM to PC ratio is 2.3-fold higher in Jensen hepatomas when compared with regenerating rat liver cells. In a variety of hepatomas the principal increase in SPM is seen in the mitochondrial and nuclear membranes which are almost SPM-free in normal cells.[46,212] It is of interest that in at least one type of hepatoma in which there is a marked increase in SPM, the level of the SPM exchange-protein is also markedly elevated.[213] A similar increase in the relative content of SPM was observed in other malignant systems such as Ehrlich ascites cells in which SPM reaches 26% of the total phospholipids.[46,47,214,215]

Senile cataract of the eye is another disease in which SPM level is elevated above the normal increase caused by aging, while the PC and PE levels are reduced. The change in lipid composition may be related to transport activity and membrane abnormality and/or to the insolubilization of proteins. The SPM of the cataract is more saturated than that isolated from normal patients. The increase in SPM content is again paralleled by increase in cholesterol.[216,217]

D. Studies on Cells in Culture

Tissue culture may serve as an excellent means for studying the contribution of membrane lipid composition to the membrane structure and function. The main advantage of such an approach is that the lipid composition can be manipulated biochemically (through lipid metabolism) or physically (by the use of liposomes or modified lipoproteins) in a controlled fashion.

Unfortunately, almost no results of studies in this direction on the role of SPM are yet available. Preliminary results obtained in our laboratory for pure cultures of myocytes and fibroblasts prepared from hearts of newborn rats are promising.[236,237] Both cell types show increase in SPM/PC and cholesterol/phospholipids mole ratio with culture age. Other changes observed in these cells with "aging" in culture include increase of the specific activities (per cell) of at least a few membrane-bound enzymes, changes in cell shape, and with the myocytes, decrease in the beating rate from 160 ± 15 beat/min (in the first week in culture) to almost none on the 13th day. Interrelationships between all these changes during cell "aging" are plausible. This is demonstrated by reversing the lipid composition of the "old" cells by lipid exchange with egg PC. After such a treatment, the lipid composition of the cells returned to that of "young" cells (regarding SPM/PC and cholesterol/phospholipids mole ratios), which is paralleled by an increase in beating rate and decrease in enzyme activities to the level of "young" cells. Plasma membrane organization and dynamics can also be related to the changes in lipid composition for both cell types.

E. Correlation with Membrane Properties

1. Membrane Integrity

Enzymatic hydrolysis of the lipid components of biological membranes might result in cell lysis. However, hydrolysis of 80% of the SPM in human erythrocytes by sphingomyelinase of *S. aureus*, and the resulting accumulation of ceramide in the membrane does not result in erythrocyte lysis.[159] Similar results were also obtained with ruminant erythrocytes in which SPM is the main phospholipid and the SPM to PC mole ratio might reach a value larger than 12[186,188] (see Table 4). This shows that under iso-osmotic conditions, the conversion of SPM to ceramide, which remains in the membranes, does not lead to loss of membrane integrity. Erythrocytes treated in this fashion are, however, osmotically fragile and after sphingomyelinase treatment, immediate lysis of ruminant erythrocyte occurs at 4°C.[218] The "cold shock" is probably due to a phase separation which occurs at low temperatures and results in membrane disintegration.

Conversion of erythrocyte SPM to ceramide also makes the membrane PC susceptible to attack by phospholipase C with resulting hemolysis.[187] A similar effect was observed with porcine and chicken erythrocytes where hydrolysis of sphingomyelin to ceramide is required for subsequent hydrolysis of PC by phospholipase D, but hemolysis does not follow.[219] In toad erythrocytes which contain small amounts of sphingomyelin, phospholipase C is able to promote hydrolysis of glycerophospholipids without prior hydrolysis by sphingomyelinase.[188] This hydrolysis leads to lysis when the cells are depleted of ATP.[188]

Another aspect of membrane integrity is its susceptibility to damage by detergents. Thus, the effect of Triton®-X-100 on lipid bilayer is related to its lipid composition.[220] It is of interest that the solubilization of phospholipids and cholesterol from erythrocyte membranes by Triton®-X-100 is differential. Of all the lipids present in this membrane, SPM and cholesterol are preferentially retained in the pellet and require much higher concentrations of detergent for their solubilization. This might indicate that in the plane of the membrane there are domains enriched in SPM and cholesterol which have a much higher apparent microviscosity[137] while the regions enriched in PC have a much lower apparent microviscosity. Because of their own low microviscosity, the Triton® molecules preferentially incorporated first to the latter regions and only subsequently interact with SPM.

Erythrocyte membranes with low SPM content are more sensitive to hemolysis by the bile salt glycocholate than erythrocytes containing high SPM content such as sheep erythrocytes.[221] Hemolysis of the latter is obtained when their PC content is increased,[222] thereby reducing their SPM to PC mole ratio.

2. Mechanical Properties and Membrane Organization

The apparent microviscosity of biological membranes as determined by DPH may be an expression of the degree of order (short-range organization) and free volume in the membrane (see Chapters 1 and 2 of this book). A positive correlation between membrane SPM content and apparent microviscosity has been shown for erythrocytes from a variety of mammalian species.[137,222] The stability of erythrocytes under iso-osmotic conditions is also greater with those richer in SPM although the resistance to osmotic shock appears to be lower.[38] We described a similar relationship with SPM content for "apparent microviscosity" and osmotic fragility in multilamellar liposomes composed of SPM and PC.[58] Recent studies in our laboratory indicate strong effect of SPM content on the free volume of lipid bilayers.[239]

The mechanical properties of the erythrocyte membranes also seem to correlate with SPM content. Cooper and co-workers[223] showed that membranes obtained from acanthocytic erythrocytes in patients with abetalipoproteinemia are enriched in SPM and depleted of PC. The SPM to PC ratio can be as high as 1.56 in acanthocytes while the normal value is about 0.86. Associated with this change is an increase in "apparent microviscosity" of the membrane and a prolonged filtration time in nucleopore filter indicative of decreased membrane deformability.[223] Recently, it was suggested that the increase of the apparent microviscosity in acanthocytes is related to the interaction of SPM with cholesterol.[154] Other examples are viral membranes which are usually richer in SPM and cholesterol than their host cell plasma membrane and therefore more ordered and closely packed than the latter.[23,24,59,180,183,224] However, the correlation of higher "apparent microviscosity" with viral infectivity[16] is questionable since the "apparent microviscosity" of the viral membrane was reduced dramatically by hydrolysis of phospholipid headgroup using phospholipase C although the effect on viral infectivity was minimal.[225] On the other hand, oxidation or removal of cholesterol which also reduces the "apparent microviscosity" reduced viral infection dramatically.[16,225] This suggests that the organization of the membrane, rather than its apparent microviscosity, is more relevant to viral infectivity.

3. Permeability and Transport

The permeability of erythrocyte membranes from a variety of mammalian species to various nonelectrolytes and electrolytes decreases with increase in SPM content; cholesterol is not an important factor in this case since its mole fraction is virtually identical in erythrocyte membranes of all mammalian species.[9,10,38] Variations in fatty acid composition have only a small effect on the erythrocyte permeability to molecules for which no specific transport system exists. A strikingly similar correlation between SPM content and the permeability of water and glucose in multilamellar liposomes has been reported by Hertz and Barenholz[58] (see also Section IV.A.1).

The water permeability of artificial membranes of SPM differs considerably from those composed of PC.[58,226] This was accounted for by the higher degree of saturation and the likelihood of hydrogen bonding in SPM bilayers.[12,58,73,226] Similar results were obtained for the permeability through the lipid bilayer of 6-carboxy fluorescein. Liposomes of bovine brain SPM showed practically no leak of this compound even at the gel-to-liquid crystalline phase transition, a consequence of its high content of nervonic acid (C24:1).[240] The contribution of the *N*-nervonyl-SPM to permeability properties of SPM is discussed in more detail in Section IV.A.1.

SPM content also affects active transport. Kirk[227] showed that the level of active transport of K^+ in various mammalian erythrocytes is inversely related to the SPM to PC ratio. It was suggested that interaction of SPM with cholesterol in the sheep erythrocyte membranes induces changes in the microenvironment of the Na^+ and K^+ pumps and is the reason for the discontinuity in the Arrhenius plots of K^+ flux in these membranes which are very enriched in SPM[228] (see also Table 4).

VII. SUMMARY AND CONCLUSIONS

Sphingomyelin is one of the major lipids of mammalian cell membranes. In most normal cells there is a gradient of SPM; its highest content is in the plasma membrane, the lowest in the inner mitochondrial membrane and the nuclear membrane which are almost free of SPM. The content of SPM in other subcellular membranes and organelles varies between these two extremes.

In cell plasma membranes, the two choline-containing lipids PC and SPM constitute more than 50% of the total phospholipids. SPM content increases with aging, especially in tissues which have a relatively low phospholipid turnover (i.e., nervous tissue and blood vessels). It also increases in several diseases, including major diseases such as atherosclerosis, certain types of cancer (e.g., hepatoma), eye cataract, and in some genetic disorders. In other diseases (such as leukemia), the SPM content is reduced. In general there is a good positive correlation between the content of SPM and cholesterol in membranes, and changes in the content of one are followed by a comparable change in the other. It is still not clear how cells maintain varying lipid composition in their different membranes. It is possible that the lipid composition of membranes is established during membrane biosynthesis and assembly. The pathological changes in SPM content might result from changes in the metabolism of this compound, i.e., increase in its rate of biosynthesis, reduction in its rate of degradation, or changes in transfer in or out of the affected cell.

It is now well established that most membrane functions are closely related to the lipid composition which affects the physical properties of the membrane. One of the main composition variables of all membranes is the SPM to PC mole ratio, while their total content is nearly constant — about half of membrane phospholipids. These two lipids are concentrated in one face of the membrane bilayer, and for plasma membranes, mainly in the outer leaflet. Variations in SPM/PC in artificial bilayers and biological membranes have a pronounced effect on the system properties. Perhaps the most striking differences between natural PC and SPM are the temperatures of the gel-to-liquid crystalline phase transition exhibited in bilayers. Most of SPMs have their transition temperatures in the physiological temperature range, while almost all naturally occurring PCs (except the dissaturated PCs of the lung surfactant) are well above their transition temperature at 37°C.

SPM and PC are classified in the same lipid subclass and have very similar packing parameters, which results from their similar inogenic group (phosphorylcholine) and their hydrophobic region (composed of two hydrocarbon chains). These gross similarities enable them to replace each other in membranes. However, there are marked dissimilarities of structure in the fine details of these molecules. PC has two hydrocarbon chains of about equal length, while SPM has one hydrocarbon chain of constant chain length, contributed by sphingosine, and a second of varying length, contributed by the *N*-acyl group. The latter can be of up to 10 carbons longer than the sphingosyl residue. This asymmetric nature is responsible in part for properties such as formation of metastable gel phase in cases where more than one gel phase is present, and the possibility of interdigitation between the two opposing monolayers of the lipid bilayer. The generally lower degree of unsaturation of SPM relative to PC is another important factor which contributes to the increase in the bilayer rigidity by SPM. A third factor is the difference in hydrogen bond-forming capability of the belt region (interface region), which connects the polar and apolar parts of these molecules. The amide bond and hydroxyl group of SPM can act as hydrogen bond donors and acceptors, while in PC the carboxyl oxygens can act only as hydrogen acceptors. The difference in the molecular structure is therefore sufficient to pose a distinct phase separation of SPM at the range of 30 to 40°C on bilayer composed of naturally occurring SPMs and PCs.

These characteristics are further reflected in the "apparent microviscosity" of mixed bilayer systems at 37°C, which increases with increasing content of SPM. Also, the phase behavior of bilayers composed of these two choline-containing lipids is strongly influenced by the addition of cholesterol. There is compelling evidence to suggest that the interaction of cholesterol with SPM is stronger than with PC. Therefore, a low mole fraction cholesterol may induce lateral phase separation of SPM-cholesterol domains in a milieu of unsaturated PC domains which are poorer in cholesterol. Thus, the microscopic phase configuration of simple bilayered systems is markedly affected by the relative concentration of SPM, PC, and cholesterol. By inference, the same situation probably exists in the bilayers of cell plasma membranes, as was demonstrated in myocyte and fibroblast cultures derived from hearts of newborn rats. The main conclusion derived from this review is that the thermotropic and phase behaviors, as well as the fluidity and permeability of membranes, are related mainly to differences in the hydrophobic region of the two choline-phospholipids. In contrast, interaction with water, with various proteins and possibly with cholesterol, relates at least in part to differences in the interface region, mainly hydrogen bonding capabilities.

Finally, manipulation of the SPM to PC ratio, both in vivo and in vitro, can be approached by passive exchange mechanisms or via metabolic pathways. This could serve as a new tool in membrane engineering which remains to be explored.

ACKNOWLEDGMENTS

The work discussed in this review was supported in part by USPHS NIH Grant number HL17576 and by US-Israel BSF Grants number 2669 and 2772. The author wishes to thank Professors S. Gatt and T. E. Thompson for the excellent discussions and involvement throughout the many years of collaborative research on the various aspects of sphingomyelin. Professor D. Shapiro is especially acknowledged for his major and pioneering contribution to the organic chemistry of sphingolipids which enabled us to carry out important parts of this research. The assistance of Mrs. Susan Nagus and Mrs. June Morris in the preparation of this manuscript is gratefully acknowledged.

126. **Lee, A. G., Birdsall, N. J. M., Levine, Y. K., and Metcalfe, J. C.,** High resolution proton relaxation studies of lecithins, *Biochim. Biophys. Acta,* 255, 43, 1972.
127. **Horowitz, A. F.,** in *Membrane Molecular Biology,* Fox, C. F. and Keith, A. D., Eds., Sinauer Associates, Sunderland, Massachusettes, 1972, 164.
128. **Seelig, A. and Seelig, J.,** The dynamic structure of fatty acyl chains in a phospholipid bilayer measured by deuterium magnetic resonance, *Biochemistry,* 13, 4839, 1974.
129. **Keough, K. M., Oldfield, E., and Chapman, D.,** *Chem. Phys. Lipids,* 10, 37, 1973.
130. **Litman, B. J. and Barenholz, Y.,** The optical activity of D-erythrocyte-sphingomyelin and its contribution to the circular dichroism of sphingomyelin-containing systems, *Biochim. Biophys. Acta,* 394, 166, 1975.
131. **Chen, G. and Kane, J. P.,** *Biochemistry,* 14, 3357, 1975.
132. **Yeagle, P. L., Hutton, W. C., Huang, C., and Martin, R. B.,** *Proc. Natl. Acad. Sci. U.S.A.,* 72, 3477, 1975.
133. **Levine, Y. K., Birdsall, N. J. M., Lee, A. G., and Metcalfe, J. C.,** *Biochemistry,* 11, 1416, 1972.
134. **Lee, A. G., Birdsall, N. J. M., and Metcalfe, J. C.,** in *Methods in Membrane Biology,* Vol. 2, Korn, E. D., Ed., Plenum Press, New York, 1974, 1.
135. **Verkleij, A. J., Zwaal, R. F. A., Roelofsen, B., Comfurius, P., Kastelijn, D., and Van Deenen, L. L. M.,** The asymmetric distribution of phospholipids in the human red cell membrane: a combined study using phospholipases and freeze-atch electron microscopy, *Biochim. Biophys. Acta,* 323, 178, 1973.
136. **Untracht, S. H., and Shipley, G. G.,** Molecular introduction between lecithin and sphingomyelin, *J. Biol. Chem.,* 252, 4449, 1977.
137. **Borochov, H., Shinitzky, M., and Barenholz, Y.,** *Cell Biophys.,* 1, 219, 1979.
138. **Sklar, L. A. and Datz, E. A.,** *Methods Enzymol.,* 81, 685, 1982.
139. **Castellino, F. J.,** *Arch. Biochem. Biophys.,* 184, 465, 1978.
140. **Berden, J. A., Cullis, P. R., Hoult, D. I., McLaughlin, A. C., Radda, G. K., and Richards, R. E.,** Frequency dependence of ^{31}P NMR linewidths in sonicated phospholipid vesicles: effects of chemical shift anisotropy, *FEBS Lett.,* 46, 55, 1974.
141. **Berden, J. A., Barker, R. W., and Radda, G. K.,** NMR studies on phospholipid bilayers: some factors affecting lipid distribution, *Biochim. Biophys. Acta,* 375, 186, 1975.
142. **Vandenheuvel, F. A.,** *J. Am. Chem. Oil Soc.,* 40, 455, 1963.
143. **Patton, S.,** *J. Theor. Biol.,* 29, 489, 1970.
144. **Oldfield, E. and Chapman, D.,** *FEBS Lett.,* 21, 303, 1972.
145. **Estep, T. N., Mountcastle, D. B., Barenholz, Y., Biltonen, R. L., and Thompson, T. E.,** *Biochemistry,* 18, 2112, 1979.
146. **Ladbrooke, B. D., Williams, R. M., and Chapman, D.,** Studies on lecithin-cholesterol-water interactions by differential scanning calorimetry and X-ray diffraction, *Biochim. Biophys. Acta,* 150, 333, 1968.
147. **Freire, E. and Snyder, B.,** Monte Carlo studies of the lateral organization of molecules in two-component lipid bilayers, *Biochim. Biophys. Acta,* 600, 643, 1980.
148. **Demel, R. A., Jansen, J. W. C. M., Van Dijck, P. W. M., and Van Deenen, L. L. M.,** The preferential interaction of cholesterol with different classes of phospholipids, *Biochim. Biophys. Acta,* 465, 1, 1977.
149. **Van Dijck, P. W. M.,** Negatively charged phospholipids and their position in the cholesterol affinity sequence, *Biochim. Biophys. Acta,* 555, 89, 1979.
150. **McCabee, P. J. and Green, C.,** *Chem. Phys. Lipids,* 20, 319, 1977.
151. **Reiber, H.,** Cholesterol-lipid interactions in membranes. The saturation concentration of cholesterol in bilayers of various lipids, *Biochim. Biophys. Acta,* 512, 72, 1978.
152. **Burns, C. H. and Rothblat, G. H.,** Cholesterol excretion by tissue culture cells: effect of serum lipids, *Biochim. Biophys. Acta,* 176, 616, 1969.
153. **Gottleib, M. H.,** The reactivity of human erythrocyte membrane cholesterol with a cholesterol oxidase, *Biochim. Biophys. Acta,* 466, 422, 1977.
154. **Barenholz, Y., Yechiel, E., Cohen, R., and Deckelbaum, R. L.,** *Cell Biophys.,* 13, 115, 1981.
155. **Pal, R., Barenholz, Y., and Wagner, R. R.,** Effect of cholesterol concentration on organization of viral and vesicle membranes, probes by accessibility to cholesterol oxidase, *J. Biol. Chem.,* 255, 5802, 1980.
156. **Barenholz, Y., Roitman, A., and Gatt, S.,** *J. Biol. Chem.,* 241, 3731, 1966.
157. **Heller, M. and Shapiro, B.,** *Biochem. J.,* 98, 763, 1966.
158. **Rachmilewitz, D., Eisenberg, S., Stein, Y., and Stein, O.,** Phospholipases in arterial tissue. I. Sphingomyelin choline phosphydrolase activity in human dog, guinea pig, rat and rabbit arteries, *Biochim. Biophys. Acta,* 144, 624, 1967.
159. **Roelofsen, B. and Zwall, R. F. A.,** in *Methods in Membrane Biology,* Vol. 7, Korn, E. D., Ed., Plenum Press, New York, 1976, 177.
160. **Bernheimer, A. W.,** Interactions between membranes and cytolytic bacterial toxins, *Biochim. Biophys. Acta,* 344, 27, 1974.
161. **Gatt, S.,** *Biochem. Biophys. Res. Commun.,* 76, 235, 1976.
162. **Yamaguchi, S. and Suzuki, K.,** *J. Biol. Chem.,* 252, 3805, 1977.

163. **Hirschfeld, T. B. and Loyter, A.,** *Arch. Biochem. Biophys.,* 67, 186, 1975.
164. **Otnasess, A. B.,** *FEBS Lett.,* 114, 202, 1980.
165. **Schneider, P. B. and Kennedy, E. P.,** *J. Lipid Res.,* 8, 202, 1967.
166. **Gatt, S., Dinur, T., and Kopolovic, J.,** *J. Neurochem.,* 31, 547, 1978.
167. **Cohen, R. and Barenholz, Y.,** Correlation between the thermotropic behavior of sphingomyelin liposomes and sphingomyelin hydrolysis by sphingomyelinase of *staphylococcus aureus, Biochim. Biophys. Acta,* 509, 181, 1978.
168. **Kramer, R., Schlatter, C., and Zahler, P.,** Preferential binding of sphingomyelin by membrane proteins of the sheep red cell, *Biochim. Biophys. Acta,* 282, 146, 1972.
169. **Widnell, C. C. and Unkless, J. C.,** *Proc. Natl. Acad. Sci. U.S.A.,* 61, 1050, 1968.
170. **Widnell, C. C.,** *Methods Enzymol.,* 32, 368, 1972.
171. **Evans, W. H. and Gurd, J. W.,** *Biochem. J.,* 133, 189, 1973.
172. **McElhaney, R. N.,** *Current Topics in Membrane Transport,* Academic Press, New York, 1983.
173. **Sood, C. K., Sweet, C., and Zull, J. E.,** Interaction of binding (Na$^+$-K$^+$) ATPase with phospholipid model membrane system, *Biochim. Biophys. Acta,* 282, 429, 1972.
174. **Linder, R., Benheimer, A. W., and Kim, K.,** Interaction between sphingomyelin and a cytolysin from the sea anemone *Stoichactis heliantus, Biochim. Biophys. Acta,* 467, 290, 1977.
175. **Watkins, M. W., Hitt, A. S., and Bulger, J. E.,** *Biochem. Biophys. Res. Commun.,* 79, 640, 1977.
176. **Herbert, P. N., Gotto, A. M., and Fredrickson, D. S.,** in *The Metabolic Basis of Inherited Disease,* 4th ed., Stanury, J. B., Wyngaarden, J. B., and Redrickson, D. S., Eds., McGraw-Hill, New York, 1978, 544.
177. **Stoffel, W., Zierenberg, O., Tunggal, B., and Schreiber, E.,** *Z. Physiol. Chem.,* 355, 1381, 1974.
178. **Arnon, R. and Teitelbaum, D.,** *Chem. Phys. Lipids,* 13, 352, 1974.
179. **Eisenberg, S. and Levy, R. I.,** *Adv. Lipid Res.,* 13, 1, 1975.
180. **Klenk, H. D.,** in *Biological Membrane,* Vol. 2, Chapman, D., Ed., Academic Press, London, 1973, 145.
181. **Patzer, E. J., Wagner, R. R., and Dubovi, E. J.,** *CRC Crit. Rev. Biochem.,* 165, 217, 1979.
182. **Lenard, J. and Compans, R. W.,** The membrane structure of lipid-containing viruses, *Biochim. Biophys. Acta,* 344, 51, 1974.
183. **Blough, H. A. and Tiffany, J. M.,** *Adv. Lipid Res.,* 11, 267, 1973.
184. **Thompson, T. E.,** in *Molecular Specialization and Symmetry in Membrane Function,* Solomon, A. K. and Karnovsky, M., Eds., Harvard University Press, Cambridge, 1978, 78.
185. **Op den Kamp, J. A. F.,** *Annu. Rev. Biochem.,* 48, 47, 1979.
186. **Zwaal, R. F. A., Roelofsen, B., and Colley, C. M.,** Localization of red cell membrane constituents, *Biochem. Biophys. Acta,* 300, 159, 1973.
187. **Zwaal, R. F. A., Roelofsen, B., Comfurius, P., and Van deenen, L. L. M.,** Organization of phospholipids in human red cell membranes as detected by the action of various purified phospholipases, *Biochim. Biophys. Acta,* 406, 83, 1975.
188. **Gazitt, Y., Ohad, I., and Loyter, A.,** Changes in phospholipid susceptibility forwards phospholipases induced by ATP depletion in avian and amphibian erythrocyte membranes, *Biochim. Biophys. Acta,* 382, 65, 1975.
189. **Bloj, B. and Zilversmit, D. B.,** *Biochemistry,* 15, 1277, 1976.
190. **Bretscher, M. S.,** Membrane structure: some general principles, *Science,* 181, 622, 1973.
191. **Gordesky, S. E., Marinetti, G. V., and Love, R.,** *J. Membr. Biol.,* 20, 111, 1975.
192. **Whiteley, N. M. and Berg, H. C.,** *J. Mol. Biol.,* 87, 541, 1974.
193. **Renooij, W., Van Golde, L. M. G., Zwaal, R. F. A., and Van Deenen, L. L. M.,** *Eur. J. Biochem.,* 61, 53, 1976.
194. **Billington, D., Coleman, R., and Tusak, Y. A.,** Topographical dissection of sheep erythrocyte membrane phospholipids by tanrocholate and glycocholate, *Biochim. Biophys. Acta,* 466, 526, 1977.
195. **Sandara, A. and Pagano, R. E.,** *Biochemistry,* 17, 332, 1978.
196. **Eisenberg, S., Stein, Y., and Stein, O.,** Phospholipase in arterial tissue. IV. The role of phosphatidyl acyl hydrolase in the regulation of phospholipid composition in the normal human aorta with age, *J. Clin. Invest.,* 48, 2320, 1969.
197. **Rouser, G. and Solomon, R. D.,** *Lipids,* 4, 232, 1969.
198. **Smith, E. B. and Cantab, B. A.,** *Lancet,* 1, 799, 1960.
199. **Eisenberg, S., Stein, Y., and Stein, O.,** Phospholipases in arterial tissue. III. Phosphatide acyl-hydrolase, lysophosphatile acyl-Hydrolase and sphingomyelin choline phosphohydrolase in rat and rabbit aorta in different age groups, *Biochim. Biophys. Acta,* 176, 557, 1969.
200. **Smith, E. B.,** *Adv. Lipid Res.,* 11, 267, 1974.
201. **Seth, S. K. and Newman, H. A. J.,** *Circ. Res.,* 36, 294, 1975.
202. **Portman, O. W., Alexander, M., and Maruff, C. A.,** *Arch. Biochem. Biophys.,* 122, 344, 1967.
203. **Portman, O. W.,** *Ann. N.Y. Acad. Sci.,* 162, 120, 1969.

According to the generally accepted fluid mosaic model of membranes, the lipid phase of the membrane is arranged in a bilayer into which proteins are embedded in varying degrees.[8] In several membrane systems studied, phospholipids were found to be asymmetrically distributed with PC primarily on the outer bilayer leaflet, while PE was located on the inner leaflet.[9,10] In studies with rat erythrocyte ghosts, Hirata and Axelrod[4] have shown that two methyltransferase enzymes are also asymmetrically distributed in membranes. It was further demonstrated that the first enzyme PMTI and its substrate PE are located on the cytoplasmic side of the membrane, whereas the second enzyme PMTII and its product PC face the outside of the membrane. PME, the monomethylated product of PMTI was found to be buried within the membrane bilayer, since it was not susceptible to digestion by phospholipase C from inside or outside of erythrocyte ghosts. These findings and further experiments suggested that the phospholipids are translocated (flip-flop) from the inside to the outside of the membranes as they are methylated.

Furthermore, when phospholipids were methylated by incorporating SAM into resealed erythrocyte ghosts, the fluorescence polarization of DPH probe decreased more than 30% as SAM concentration was increased.[11] This increase in membrane fluidity was completely eliminated when *S*-adenosyl-L-homocysteine (SAH) was present inside the cells together with SAM. SAH inhibited methylation of PE and did not affect the membrane fluidity by itself.[11,12] In experiments with varying concentrations of methyl-^{3}H labeled SAM, it was found that the membrane fluidity reached almost maximum at a concentration of SAM about 25 μM. From thin layer chromatography of the methylated products it was confirmed that at SAM concentration below 25 μM, PME was the major component, whereas the synthesis of PC increased at SAM concentration above 25 μM.[3] These findings indicated that the increase in erythrocyte membrane fluidity was primarily due to the methylation of PE to PME, which accumulated mainly within the lipid bilayer.[11] Experiments with model membranes further indicated that the membrane fluidity of synthetic liposomes containing PE, PME, and PC increased when the proportion of PME was increased or the proportion of PC was decreased.[13]

In vivo treatments of rats with microsomal enzyme-inducing agents, phenobarbital and 3-methylcholanthrene significantly increased the proportion of PME, but decreased the proportion of PC during phospholipid methylation of liver microsomal membranes by low and high concentration of SAM.[13] In addition, administration of both of these compounds increased membrane fluidity as measured by fluorescence polarization of DPH probe.[13]

The combined results of these experiments strongly suggested a close association between changes in membrane fluidity and the phospholipid methylation process in vitro as well as in vivo systems.

III. MODULATION OF β-ADRENERGIC RECEPTOR SYSTEM

Reticulocytes (immature red cells) contain β-adrenergic receptors that can bind to agonists and subsequently can couple with adenylate cyclase to generate cyclic AMP.[14] Axelrod and co-workers[15,16] have demonstrated that when [methyl-^{3}H] SAM was introduced inside rat reticulocyte ghosts, incorporation of [^{3}H] methyl group into phospholipids increased significantly after the ghosts were exposed to β-adrenergic agonist L-isoproterenol. Such stimulation of phospholipid methylation by L-isoproterenol was found to be stereospecific, since dextro isomer of isoproterenol was ineffective. The concentration of L-isoproterenol for half maximal activation of adenylate cyclase was close to that required for half maximal activation of lipid methylation. The increase in phospholipid methylation by a number of catecholamine agonists followed the same order of potency as the activation of adenylate cyclase which was characteristic of the β-adrenergic receptor. GTP enhanced the isoproterenol-stimulated phospholipid methylation in a manner similar to its ability to increase the adenylate cyclase activity in rat reticulocytes. Both phospholipid methylation and cyclic AMP formation were inhibited by the β-adrenergic antagonist, propranolol, but not by phentolamine, an α-adrenergic antagonist. Direct stimulation of adenylate cyclase with cholera toxin or sodium fluoride, which bypasses the receptor, did not increase the phospholipid methylation. All of these findings suggested that the membrane phospholipid methylation increased due to the binding of an agonist to the β-adrenergic receptor and not as a consequence of adenylate cyclase activation.

From the fluorescence polarization measurements of DPH probe, it was further indicated that the phospholipid methylation increased the fluidity of the reticulocyte membranes.[16] Previous studies have shown that the addition of *cis*-vaccenic acid to turkey erythrocyte ghosts increased the membrane fluidity as measured by the fluorescence polarization of DPH and enhanced the coupling between hormone receptors and adenylate cyclase.[17] Other workers have also reported that the modulation of membrane fluidity by various agents produced similar changes in the adenylate cyclase activity.[18-20]

The results from these experiments indicated that as the phospholipid methylation is increased due to the binding of an agonist to the receptor, the membrane fluidity is increased, and the greater lateral mobility of the hormone receptor facilitates its coupling with adenylate cyclase.[16]

IV. LYMPHOCYTE MITOGENESIS

The binding of a variety of lectins (plant proteins), concanavalin A (Con A) and phytohemagglutinin (PHA) to lymphocytes is known to trigger the stimulation of DNA synthesis and mitogenesis.[21] The mechanism of such cellular activation processes has been studied extensively as a model for understanding the more general problem of regulation of cell proliferation. Recent studies have shown that the membrane binding of lectins and the mobility of receptors on cell surfaces are important factors in lymphocyte mitogenesis.[22]

Using the spin label probe sodium 6 (4′,4-dimethyl-oxazolidinyl-*N*-oxyl)-heptadecanoate, Barnett et al.[23] reported a decrease in ordering in membranes of human peripheral lymphocytes treated with PHA and also in membranes of mouse spleenic lymphocytes treated with Con A. A maximum membrane fluidity was observed between 15 and 30 min of mitogenic lectin treatment and the dose response curve of the membrane fluidization corresponded well with that for mitogenesis. In contrast, a nonmitogenic lectin, wheat germ agglutinin, did not have any effect on lymphocyte membrane fluidity. In another study, Curtain and co-workers[24] reported that mitogenic agents, Con A, PHA, Calcium ionophore A 23187, and periodate caused a 20% decrease in lipid ordering in the bilayer region of lymphocyte plasma membranes, as measured by 5-nitroxide stearic acid spin label. Fluorescence polarization measurements of DPH indicated that the mitogenic lectins, *W. floribunda* mitogen and *Lens culinaris* hemagglutinin, increased membrane fluidity upon binding to human peripheral lymphocytes within 30 min, whereas the nonmitogenic lectins did not affect the membrane fluidity.[25]

Recently, Hirata et al.[5] have shown that the stimulation with mitogenic concentration of Con A caused about a doubling in phospholipid methylation and also a two- to threefold increase in the accumulation of [³H]-lysophosphatidylcholine in murine spleen lymphocytes incubated with [³H] methionine. This increase in methylation reached a maximum at 10 min and then returned to control levels in about 40 min with a concomitant activation of phospholipase A_2. While the dose response curves of Con A for phospholipid methylation and thymidine incorporation in lymphocytes were almost parallel, nonmitogenic lectins did not affect the methylation of phospholipids. A close association between phospholipid methylation and mitogenesis was further established by the fact that the inhibition of either methylation or phospholipase A_2 resulted in a decrease in the thymidine incorporation.

Working with chicken erythrocytes, Nakajima et al.[26] reported similar stimulation of phospholipid methylation immediately after the binding with Con A. Using spin labeled fatty acids, these workers measured the membrane fluidity changes in erythrocytes at various times after exposure of Con A.

Electron spin resonance spectra showed changes in the peaks due to the spin probes in membrane lipids and indicated an increase in membrane fluidity which could occur in three phases. The first increase in membrane fluidity, which was observed within 10 min of Con A binding could be blocked by a methyl transferase inhibitor, S-isobutyryl-adenosine (SIBA) and hence was temporally related to phospholipid methylation. While the second phase of fluidity increase was prevented by cytochalasin B, which disrupts microfilaments, the last phase was inhibited by the Ca^{2+}-chelating reagent, EGTA. Since Ca^{2+} uptake was gradually increased after 10 min of addition of Con A and the calcium influx was stimulated by phospholipid methylation, it was suggested that the final phase of membrane fluidity could be regulated by both Ca^{2+} and phospholipid methylation. The results of these experiments indicated that phospholipid methylation increases membrane fluidity which then facilitates other membrane-associated events and subsequently lymphocyte mitogenesis.

V. IN VIVO TREATMENT WITH S-ADENOSYLMETHIONINE

A. Ethinyl Estradiol-Induced Cholestasis

Several drugs including estrogens, anabolic steroids, and phenothiazines have been shown to produce intrahepatic cholestasis (bile secretory failure) in humans and experimental animals.[27] Ethinyl estradiol (EE), a synthetic estrogen, has been extensively studied to delineate the mechanisms of its pharmacological and toxicological actions.[28] Clinical studies have confirmed that the administration of EE produces reduced bile flow, supersaturation of bile with cholesterol, and decreased capacity to excrete organic anions, which might explain in part the increased risk of gall bladder diseases in women using oral contraceptives that contain EE.[29,30]

A daily dose of EE (5 mg/kg body wt) to rats for 5 days decreases bile flow and hepatic NaK-ATPase to 50%, and increases free cholesterol (130%) and cholesterol ester (400%) with no change in total phospholipid content of liver plasma membranes (LPM).[31]

Electron spin resonance studies of such EE-treated rat LPM indicated that the rotational correlation time of isotropic spin probe methyl (12-nitroxyl) stearate increased 60% and the order parameter, measured by anisotropic spin label (5-nitroxide) stearic acid increased 5%, thereby demonstrating a significant immobilization of these spin probes. Studies with [^{3}H]ouabain and [γ^{32}P]ATP binding to LPM indicated no change in the number of enzyme units of NaK-ATPase in EE-treated rats and hence the reduction of NaK-ATPase activity was attributable to the changes in the membrane structure caused by decreased LPM fluidity.[31]

A uniform correlation between EE-induced decrease in LPM NaK-ATPase activity and membrane fluidity as measured by fluorescence polarization of DPH was also reported by Keefe and co-workers.[32]

Recently, Stramentinoli and co-workers[33-35] have shown that simultaneous administration of EE and disulfate-di-p-toluene sulfonate, a stable salt of SAM, offered protection against EE-induced cholestasis and the bile flow was restored to its normal level. Furthermore, (methyl ^{14}C) SAM when given orally was rapidly taken up by the liver without prior chemical modification and the radioactivity associated with PC extracted from LPM increased with time.[36] These experiments indicated that exogenous SAM acts as a methyl donor in the formation of PC from PE. Further studies indicated that in vivo administration of SAM alone and with EE increased LPM PC content 30%.[37] In contrast, SAM did not alter cholesterol level of rat LPM by either treatment.[37]

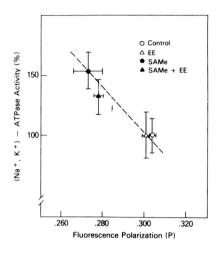

FIGURE 1.   Relationship between NaK-ATP-
ase activity and fluorescence polarization of DPH
in isolated LPMs from female Sprague-Dawley
rats treated with SAM (25 mg/kg/day, 3 times
per day, for 3 days), EE (5 mg/kg/day, for 3
days), and the combination of the two drugs (SAM
+ EE). Mean ± S.E. (four to seven rats).

Recent studies in our laboratory have demonstrated for the first time that in vivo treatment
of SAM alone increased LPM NaK-ATPase activity 51% and in combination with EE
increased the enzyme activity 31%.[38] From the fluorescence polarization measurement of
DPH, it was further established that the SAM treatment to rats significantly increased the
membrane fluidity in a fashion parallel to the increase in NaK-ATPase activity (Figure 1).[38]
We also demonstrated that the treatment of EE for a shorter period (three days) despite
inducing cholestasis did not alter LPM fluidity or NaK-ATPase activity (Figure 1) in contrast
to their decrease after five days of EE treatment.[31,32] Earlier, Simon and co-workers[39] had
reported that phenobarbital when administered to EE-treated rats restored the NaK-ATPase
activity to control levels but did not correct the decreased bile salt excretion. These combined
results suggest that pathomechanisms other than that resulting from impaired NaK-ATPase
activity are responsible for EE-induced bile secretory failure, and can be positively influenced
by phospholipid methylating reaction system.[38,39]

A parallel correlation between NaK-ATPase activity and membrane fluidity was also
observed after in vivo treatment with other agents that caused a decrease (EE) or an increase
(propylene glycol, thyroid hormone, and cortisone).[31,32] While EE administration increased
both cholesterol and cholesterol:phospholipid ratio, propylene glycol, thyroid, and cortisone
acetate treatments decreased only cholesterol level and did not alter cholesterol:phospholipid
ratio.[32] Thus, changes in the LPM fluidity produced by these agents appear to be causally
related to their abilities to alter the cholesterol levels in the membrane. In contrast, SAM
had little effect on cholesterol level and hence the increased fluidity in LPM observed after
SAM treatment was attributed to the increased phospholipid methylation.[37,38]

In contrast to NaK-ATPase, other membrane-bound enzymes, e.g., Mg-ATPase and γ-
glutamyl transpeptidase (γ-GT) did not exhibit any changes in their activities after in vivo
administration of SAM alone, indicating the specificity of SAM-mediated interactions.[38]
Furthermore, the activities of these enzymes could not be correlated with fluidity increases
produced by SAM administration. Since NaK-ATPase has been shown to be primarily
localized in a region separate from the parts of the LPM rich in Mg-ATPase and γ-GT,[40,41]
it is possible that the localized changes in membrane fluidity and lipid composition induced

by SAM could very well trigger differential responses in the enzymes embedded in different parts of the lipid matrix.[38,42,43] In agreement with these observations, in vivo administration of EE and phenobarbital to rat LPM did not show any consistent relationship between LPM fluidity and enzymatic activities of Mg-ATPase and 5'-nucleotidase, further demonstrating the selective nature of membrane perturbations caused by these agents.[32,38,39]

B. Aging

Aging produces many changes in the brain including reduced uptake of catecholamines, decrease in dopaminergic and β-adrenergic receptors, and a decline in the tissue levels of SAM.[44-47]

Recently, Algeri and co-workers[48] have shown that when [methyl-^{3}H] SAM was injected intracranially into the rat brain, in vivo incorporation of [^{3}H] methyl group into the brain PC was significantly reduced in 30-month-old rats compared to 3-month-old rats. This finding indicated a decreased synthesis of PC from PE in the aging rats. In order to detect any relationship between phospholipid methylation, membrane fluidity, and the adrenergic and dopaminergic receptor activities, these parameters were measured in the brain before and after chronic administration of SAM (50 mg/kg/day for 3 months) to senescent rats.

Fluorescence polarization measurements of DPH and electron spin resonance experiments with two spin labels 5-NS (5-ketostearic acid nitroxide) and 16-NS (16-ketostearic acid nitroxide) which probes the fluidity of the surface and the hydrophobic core of the membrane, respectively, indicated a significant reduction in the striatal membrane fluidity in the aging rats. Chronic treatment of SAM restored the fluorescence polarization to normal values. The spin probe studies demonstrated that the SAM treatment increased membrane fluidity of the hydrophobic core region with no change at the membrane surface. β-adrenergic binding sites, reduced in the aging rats were also increased upon SAM treatment. These phenomena corresponded well with Axelrod's observation that increased phospholipid methylation increased membrane fluidity and β-adrenergic binding sites in rat reticulocyte ghosts.[16,49]

Since dopaminergic binding sites in the rat striatum and the membrane fluidity in the cortex region of the brain were not affected by SAM treatment, the effects of phospholipid methylation on these properties were considered to be specific for certain brain regions and receptor systems.[48]

VI. CONCLUSION

Increased phospholipid methylation has been shown to increase membrane fluidity and enhance certain receptor-mediated interactions in several systems. However, the exact mechanism of how a small amount of methylated phospholipid can influence the membrane fluidity is not yet understood and requires further studies. It should be noted that not all receptor-associated functions involve phospholipid methylation. Stimulation of platelets with thrombin, prostaglandins, or epinephrine did not have any effect on phospholipid methylation.[16,50] In another study, phospholipid methylation was actually decreased by stimulation of chemotactic receptors in neutrophils and macrophages.[51] Examples have also been cited in the text where phospholipid methylation did not produce any changes in the membrane fluidity.

From these studies and others not described in this communication,[52,53] it can be concluded that the membrane structure-function changes modulated by phospholipid methylation appear to be highly specific with respect to the particular receptor and cell systems.

ACKNOWLEDGMENTS

I would like to thank Drs. U. A. Boelsterli, N. Tandon, E. O. Titus, M. Shinitzky, F.

Hirata, and J. Axelrod for their helpful discussions. I am very grateful to Drs. S. Algeri and G. Stramentinoli for their interest and permission to include some of their work before its publication. The typing assistance of Ms. Olivia Hoffman is gratefully acknowledged.

REFERENCES

1. **Bremmer, K. and Greenberg, D. M.,** Methyl transferring enzyme system of microsomes in the biosynthesis of lecithin (phosphatidylcholine), *Biochim. Biophys. Acta,* 46, 205, 1961.
2. **Scarborough, G. A. and Nyc, J. F.,** Methylation of ethanolamine phosphatides by microsomes from normal and mutant strains of *Neurospora crassa, J. Biol. Chem.,* 242, 238, 1967.
3. **Hirata, F., Viveros, O. H., Diliberto, E. M., Jr., and Axelrod, J.,** Identification and properties of two methyltransferases in the conversion of phosphatidylethanolamine to phosphatidylcholine, *Proc. Natl. Acad. Sci. U.S.A.,* 75, 1718, 1978.
4. **Hirata, F. and Axelrod, J.,** Enzymatic synthesis and rapid translocation of phosphatidylcholine by two methyltransferases in erythrocyte membranes, *Proc. Natl. Acad. Sci. U.S.A.,* 75, 2348, 1978.
5. **Hirata, F., Toyoshima, S., Axelrod, J., and Waxdal, M. J.,** Phospholipid methylation: a biochemical signal modulating lymphocyte mitogenesis, *Proc. Natl. Acad. Sci. U.S.A.,* 77, 862, 1980.
6. **Sastry, B. V. R., Statham, C. N., and Axelrod, J.,** Properties of two phosphatidylcholine in the rat liver, *Fed. Proc. Fed. Am. Soc. Exp. Biol.,* 38, 544, 1979.
7. **Crews, F. T., Hirata, F., and Axelrod, J.,** Identification and properties of two methyltransferases that synthesize phosphatidylcholine in rat brain synaptosomes, *J. Neurochem.,* 34, 1491, 1980.
8. **Singer, J. J. and Nicholson, G. L.,** The fluid mosaic model of the structure of cell membranes; cell membranes are viewed as two-dimensional solutions of oriented globular proteins and lipids, *Science,* 175, 720, 1972.
9. **Chap, H. J., Zwaal, R. F. A., and Van Deenen, L. L. M.,** Action of highly purified phospholipases on blood platelets; evidence for an asymmetric distribution of phospholipids in the surface membrane, *Biochim. Biophys. Acta,* 467, 146, 1977.
10. **Rothman, J. E. and Lenard, J.,** Membrane asymmetry: the nature of membrane asymmetry provides clues to the puzzle of how membranes are assembled, *Science,* 195, 743, 1977.
11. **Hirata, F. and Axelrod, J.,** Enzymatic methylation of phosphatidylethanolamine increases erythrocyte membrane fluidity, *Nature (London),* 275, 219, 1978.
12. **Cantoni, G. L., Richard, H. H., and Chiang, P. K.,** Inhibitors of *S*-adenosylhomocystein hydrolase and their role in the regulation of biological methylation, in *Transmethylation,* Usdin, E., Borchardt, R. T., and Creveling, C. R., Eds., Elsevier, Amsterdam, 1979, 155.
13. **Rama Sastry, B. V., Statham, C. N., Meeks, R. G., and Axelrod, J.,** Changes in phospholipid methyltransferases and membrane microviscosity during induction of rat liver microsomal cytochrome P-450 by phenobarbital and 3-methylcholanthrene, *Pharmacology,* 23, 211, 1981.
14. **Bilezikian, J. P., Spiegel, A. M., Gammon, D. E., and Aurbach, G. D.,** The role of guanyl nucleotides in the expression of catecholamine-responsive adenylate cyclase during maturation of the rat reticulocyte, *Mol. Pharmacol.,* 13, 786, 1977.
15. **Hirata, F., Strittmatter, W. J., and Axelrod, J.,** *Beta*-adrenergic receptor agonists increase phospholipid methylation, membrane fluidity and *beta*-adrenergic receptor-adenylate cyclase coupling, *Proc. Natl. Acad. Sci. U.S.A.,* 76, 368, 1979.
16. **Hirata, F. and Axelrod, J.,** Phospholipid methylation and biological signal transmission, *Science,* 209, 1082, 1980.
17. **Rimon, G., Hanski, E., Braun, S., and Levitzki, A.,** Mode of coupling between hormone receptors and adenylate cyclase elucidated by modulation of membrane fluidity, *Nature (London),* 276, 394, 1978.
18. **Houslay, M. D., Hesketh, T. R., Smith, G. A., Warren, G. B., and Metcalfe, J. C.,** The lipid environment of the glucagon receptor regulates adenylate cyclase activity, *Biochim. Biophys. Acta,* 436, 495, 1976.
19. **Klein, I., Moore, L., and Pastan, I.,** Effect of liposomes containing cholesterol on adenylate cyclase activity of cultured mammalian fibroblasts, *Biochim. Biophys. Acta,* 506, 42, 1978.
20. **Orly, O. and Schramm, M.,** Fatty acids as modulators of membrane functions: catecholamine-activated adenylate cyclase of the turkey erythrocyte, *Proc. Nat. Acad. Sci. U.S.A.,* 72, 3433, 1975.
21. **Oppenheim, J. J. and Rosenstreich, D. L.,** *Mitogens in Immunobiology,* Academic Press, New York, 1976, 85.
22. **Edelman, G. M.,** Surface modulation in cell recognition and growth; some new hypotheses on phenotypic alteration and transmembranous control of cell surface receptors, *Science,* 94, 218, 1976.
23. **Barnett, E. R., Scott, R. E., Furcht, L. T., and Kersey, J. H.,** Evidence that mitogenic lectins induce changes in lymphocyte membrane fluidity, *Nature (London),* 249, 465, 1974.

24. **Curtain, C. C., Looney, F. D., Marchalonis, J. J., and Raison, J. K.,** Changes in lipid ordering and state of aggregation in lymphocyte plasma membranes after exposure to mitogens, *J. Membr. Biol.*, 44, 211, 1978.

25. **Toyoshima, S. and Osawa, T.,** Lectins from *Wisteria floribunda* seeds and their effects on membrane fluidity of human peripheral lymphocytes, *J. Biol. Chem.*, 250, 1655, 1975.

26. **Nakajima, M., Tamura, E., Irimura, T., Toyoshima, S., Hirano, H., and Osawa, T.,** Mechanism of the concanavalin A-induced change of membrane fluidity of chicken erythrocytes, *J. Biochem.*, 89, 665, 1981.

27. **Plaa, G. L. and Pristley, B. G.,** Intrahepatic cholestasis induced by drugs and chemicals, *Pharmacol. Rev.*, 28, 207, 1977.

28. **Kellie, A. E.,** The pharmacology of the estrogens, *Annu. Rev. Pharmacol. Toxicol.*, 11, 97, 1971.

29. **Bennion, L. J., Mott, D. M., and Howard, B. V.,** Oral contraceptives raise the cholesterol saturation of bile by increasing biliary cholesterol secretion, *Metabolism*, 29, 18, 1980.

30. **DiPadova, C., Tritapepe, R., Cammareri, G., Humpel, M., and Stramentinoli, G.,** S-adenosyl-L-methionine antagonizes ethinyl estradiol-induced bile cholesterol supersaturation in humans without modifying the estrogen plasma kinetics, *Gastroenterology*, 82, 223, 1982.

31. **Davis, R. A., Kern, F., Jr., Showalter, R., Sutherland, E., Sinensky, M., and Simon, F.,** Alterations of hepatic Na^+, K^+-ATPase and bile flow by estrogen: effects on liver surface membrane lipid structure and function, *Proc. Natl. Acad. Sci.*, 75, 4130, 1978.

32. **Keefe, E. B., Scharschmidt, B. F., Blankenship, N. M., and Ockner, R. K.,** Studies of relationships among bile flow, liver plasma membrane NaK-ATPase, and membrane microviscosity in the rat, *J. Clin. Invest.*, 64, 1590, 1979.

33. **Stramentinoli, G., Gualano, M., and Di Padova, C.,** Effect of S-adenosyl-L-methionine on ethynylestradiol-induced impairment of bile flow in female rats, *Experientia*, 33, 1361, 1977.

34. **Stramentinoli, G., Gualano, M., Rovagnati, P., and DiPadova, C.,** Influence of S-adenosyl-L-methionine on irreversible binding of ethynylestradiol to rat liver microsomes and its implication in bile secretion, *Biochem. Pharmacol.*, 28, 981, 1979.

35. **Stramentinoli, G., Di Padova, C., Gualano, M., Rovagnati, P., and Galli-Kienle, M.,** Ethynylestradiol-induced impairment of bile secretion in the rat: protective effects of S-adenosyl-L-methionine and its implication in estrogen metabolism, *Gastroenterology*, 80, 154, 1981.

36. **Stramentinoli, G., Gualano, M., and Galli-Kienle, M.,** Intestinal absorption of S-adenosyl-L-methionine, *J. Pharmacol. Exp. Ther.*, 209, 323, 1979.

37. **Schreiber, A. J., Sutherland, E., and Simon, F. R.,** Prevention of ethinyl estradiol (EE) cholestasis by S-adenosyl-L-methionine (SAM): possible role of phosphatidyl choline (PC), *Hepatology*, 2, 697, 1982.

38. **Boelsterli, U. A., Rakhit, G., and Balazs, T.,** Modulation by S-adenosyl-L-methionine of hepatic Na^+, K^+-ATPase, membrane fluidity, and bile flow in rats with ethinyl estradiol-induced cholestasis, *Hepatology*, 3, 12, 1983.

39. **Simon, F. R., Gonzalaz, M., Sutherland, E., Accatino, L., and Davis, R. A.,** Reversal of ethinyl estradiol-induced bile secretory failure with Triton WR-1339, *J. Clin. Invest.*, 65, 851, 1980.

40. **Latham, P. S. and Kashgarian, M.,** The ultrastructural localization of transport ATPase in the rat liver at nonbile canalicular plasma membranes, *Gastroenterology*, 76, 988, 1979.

41. **Blitzer, B. L. and Boyer, J. L.,** Cytochemical localization of Na^+, K^+-ATPase in the rat hepatocyte, *J. Clin. Invest.*, 62, 1104, 1978.

42. **Kremmer, T., Wisher, M. H., and Evans, W. H.,** The lipid composition of plasma membrane subfractions originating from the three major functional domains of the rat hepatocyte cell surface, *Biochim. Biophys. Acta*, 455, 655, 1976.

43. **Evans, W. H.,** A biochemical dissection of the functional polarity of the plasma membrane of the hepatocyte, *Biochim. Biophys. Acta*, 604, 27, 1980.

44. **Jonec, V. J. and Finch, C. E.,** Senescence and dopamine uptake by subcellular fractions of the C57BL/6J male mouse brain, *Brain Res.*, 91, 197, 1975.

45. **Misra, C. H., Shelat, H. S., and Smith, R. C.,** Effect of age on adrenergic and dopaminergic receptor binding in rat brain, *Life Sci.*, 27, 521, 1980.

46. **Severson, J. A. and Finch, C. E.,** Reduced dopaminergic binding during aging in the rodent striatum, *Brain Res.*, 192, 147, 1980.

47. **Stramentinoli, G., Gualano, M., Catto, E., and Algeri, S.,** Tissue levels of S-adenosylmethionine in aging rats, *J. Gerontol.*, 32, 392, 1977.

48. **Cimino, M., Curatola, G., Pezzoli, C., Stramentinoli, G., Vantini, G., and Algeri, S.,** Age-related modification of dopaminergic and β-adrenergic receptor system: restoration of normal activity by modifying membrane fluidity with S-adenosyl methionine, in *Aging*, Vol. 23, *Aging Brain and Ergot Alkaloids*, Agnoli, A., Ed., Raven Press, New York, 1983.

49. **Strittmatter, W. J., Hirata, F., and Axelrod, J.,** Phospholipid methylation unmasks cryptic β-adrenergic receptors in rat reticulocytes, *Science,* 204, 1205, 1979.
50. **Randon, J., Lecompte, T., Chignard, M., Siess, W., Marlas, G., Dray, F., and Vargraftig, B. B.,** Dissociation of platelet activation from phospholipid methylation, *Nature (London),* 293, 660, 1981.
51. **Hirata, F.,** Overviews on phospholipid methylation, in *Biochemistry of S-adenosylmethionine and Related Compounds,* Usdin, E., Borchardt, R. T., and Creveling, C. R., Eds., Macmillan, London, 1982, 109.
52. **Collard, O. and Breton, M.,** Rat liver plasma membrane phospholipids methylation: its absence of direct relationship to adenylate cyclase activities, *Biochem. Biophys. Res. Commun.,* 101, 727, 1981.
53. **Moore, J. P., Smith, G. A., Hesketh, T. R., and Metcalfe, J. C.,** Early increases in phospholipid methylation are not necessary for the mitogenic stimulation of lymphocytes, *J. Biol. Chem.,* 257, 8183, 1982.

INDEX

A

ACAT, see Acyl CoA: cholesterol acyl transferase
Acceptor, cholesterol, 89
Accessibility, membrane receptors, 26—31, 36
    modulation of, 26—29
Acetylcholine, 28, 30
Acetylcholinesterase, 90, 161
Acetyl-CoA, 101—102
Acetyl-CoA carboxylase, 101—102
AchE, see Acetylcholinesterase
Acholeplasma laidlawii, 103
Acidity, effects of, 13
ACP, see Acyl carrier protein
ACTH, see Adrenocorticotropic hormone
Activation energy, see also Flow activation energy,
    155
Active lipid, 37—38
Active modulation, cellular function studies, 2, 24,
    31—32
Active regulation, membrane function, 2
Activity, overt, see Overt activity
Acyl-ACP, 120
Acyl carrier protein, 101—102, 106—107, 109—
    110, 118—121
Acyl chains
    collective distribution, alteration of, 123—124
    fatty acid regulation studies, 100—101, 111—125
    length, 139, 141—144
    modifications of, 112—125
        temporal relations involving, 124—125
    phospholipids, 74—75, 81, 100—101, 111—125
    specific positioning during *de novo* phospholipid
        biosynthesis, 111—112
    sphingomyelin studies, 133, 135, 137, 139—150,
        152, 155, 166
    turnover, 112—122
Acyl-CoA, 103, 106—111, 113, 118—121
    metabolic origins of, 118—119
Acyl-CoA:1-acyl-*sn*-glycerol 3-phosphate acyltrans-
    ferase, 111
Acyl CoA: cholesterol acyltransferase, 32, 86
Acyl-CoA-*sn*-glycerol 3-phosphate acyltransferase,
    111
Acyl-CoA hydrolase, 103, 106
Acyl-CoA intermediates, 102
Acyl-CoA ligase, 103
Acyl-CoA: lysophospholipid acyltransferase, 119
Acyl-CoA synthetase, 118—121
1-Acyl-dihydroxyacetone phosphate, 111
1-Acylglycerol 3-phosphate, 111
1-Acylglycerol phosphate acyltransferase, 118
N-Acyl-sphingosine-1-phosphorylcholine, see
    Sphingomyelin
Acylthioesterase, 119
Acyl transfer, 117
Acyltransferase, 86, 106, 111, 115, 118—120, 122
Adaptation, effects of, 14—16, 36

homeoviscous, 15, 133
Addiction, see Drugs, addiction to
Adenosine monophosphate, cyclic, see Cyclic AMP
Adenosine triphosphatase, 32, 90, 161
Adenosine triphosphatase, NaK-, 179—181
Adenosine triphosphate, 31, 164
S-Adenosylmethionine, 176—177
    in vivo treatment with, 179—181
Adenylate cyclase, 17, 24, 31—32, 177—178
β-Adrenergic receptor, 24, 27—29
β-Adrenergic receptor system, modulation of, in
        phospholipid methylation, 177—178, 181
Adrenocorticotropic hormone, 28, 30
Aerobic organisms, desaturation in, 106—111
Aggregation, 17—19, 21, 29—30, 63—64, 91
Aging, effects of
    cellular function studies, 6, 14—16, 27—28, 31,
        35—37
    phospholipid methylation studies, 181
    sphingomyelin studies, 132—134, 162—163, 165
Agonist-receptor binding, in phospholipid methyla-
    tion, 177—178
Alcohol, 11, 15, 37
Algae, 108—108
Aliphatic hydrocarbons, 8—9, 13, 57
Alteration mechanism, phospholipid deacylation-rea-
    cylation as, 121—122
Amino acid transport, 31, 33
Aminoisobutyric acid, 23—24
AMP, cyclic, see Cyclic AMP
Amphipathic lipid, 18, 79
Amphipathic protein, 108
Amphipaths, 150
    swelling, 134—135
Amphiphiles, 65
Anaerobic bacteria, desaturation in, 104—106
Anesthetics, effects of, 15, 37, 109, 121—122
Angiotensin II, 28
Angular coordinates, rigid body, 54—55
Animal cells, desaturation in, 108—109
Animal systems, see also Mammalian systems
    cellular function studies, 8, 15, 31
    cholesterol regulation studies, 74—91
    fatty and regulation studies, 101—102, 107—110
Anisotropy, see also Fluorescence anisotropy, 4,
        16—17, 54, 59, 62, 152, 155
Annulus, lipid, 7, 90
Antibody specificity, choline phospholipids, 161
Antigen, see also specific types by name
    cellular function studies, 18—20, 25—26, 31—
        37
    established, 34—35
    passive modulation of, 31—36
        clinical implications, 36
    physical aspect studies, 66
θ-Antigen, 34—35
Apolipoprotein, 76—78, 83—84, 88—89, 161
    distribution and function, in human plasma lipo-
        protein, 77—78